Lecture Notes in Control and Information Sciences

Edited by A. V. Balakrishnan and M. Thoma

Lecture Notes in
Control and
Information Sciences
Edited by A.V. Balakrishnan and M. Thoma

15

Semi-Infinite Programming

Proceedings of a Workshop
Bad Honnef August 30 – September 1, 1978

Edited by R. Hettich

Springer-Verlag
Berlin Heidelberg GmbH 1979

Editors
Prof. Dr. R. Hettich, Bonn

ISBN 978-3-540-09479-1 ISBN 978-3-540-35213-6 (eBook)
DOI 10.1007/978-3-540-35213-6

Originally published by Springer-Verlag Berlin Heidelberg New York in 1979

2060/3020-543210

Preface

This book contains the proceedings of a Workshop on Semi-Infinite Programming which was held in the "Honnefer Haus" of the Elly-Hölterhoff-Böcking Stiftung, Bad Honnef, August 30 - September 1, 1978, and which was organized within the scope of the activities of the Sonderforschungsbereich 72 at the University of Bonn, sponsored by the Deutsche Forschungsgemeinschaft.

Aim of the workshop was to get an impression of the state of the art in the field of semi-infinite programming with special emphasis on the practical aspects, i.e. numerical methods and applications. I thank all participants for their contributions to discussions and to these proceedings. Furthermore, I thank I. Kreuder for her assistance in organizing the workshop and the Springer-Verlag for the good cooperation.

Bonn, March 1979

Rainer Hettich

TABLE OF CONTENTS

THEORY

METHODS FOR LINEAR PROBLEMS

METHODS FOR NONLINEAR PROBLEMS

APPLICATIONS

Papers presented at the workshop but not received for the Proceedings:

Eckhardt, U.
Semi-Infinite Systems of Linear Inequalities

Hofmann, K.H.
Methods of Approximation Theory in Semi-Infinite Programming and Applications

Klostermaier, A.
A Semi-Infinite Linear Programming Procedure

LIST OF PARTICIPANTS

A. BEN-Tal	Technion - Israel Institute of Technology - Computer Science Department Haifa, Israel
H.G. BOCK	Institut f. Angewandte Mathematik, SFB 72, Universität Bonn, Wegelerstr. 6 5300 Bonn
D. BRAESS	Institut für Mathematik, Ruhruniversität Universitätsstr. 150, 4630 Bochum
U. ECKHARDT	Inst. f. Angew. Math. der Universität Bundesstr. 55, D-2000 Hamburg 13
K. GLASHOFF	Inst. f. Angew. Math. der Universität Bundesstr. 55, D-2000 Hamburg 13
G.H. GOLUB	Serra House, Serra St. Computer Science Dept. Stanford University Stanford, C.A. 94305, USA
P.R. GRIBIK	Department of Mathematics, Carnegie-Mellon University Pittsburgh, P.A. 15213, USA
S.A. GUSTAFSON	The Royal Institute of Technology, Numerical Analysis S-10044 Stockholm 70, Schweden
R. HETTICH	Institut f. Angewandte Mathematik, SFB 72, Universität Bonn, Wegelerstr. 6, 5300 Bonn
K.H. HOFFMANN	Freie Universität Berlin Institut für Mathematik III Arnimallee 2-6, 1000 Berlin 33
W. VAN HONSTEDE	Institut f. Angewandte Mathematik, SFB 72, Universität Bonn, Wegelerstr. 6, 5300 Bonn
A. KLOSTERMAIER	Freie Universität Berlin Institut für Mathematik III Arnimallee 2-6, 1000 Berlin 33
H.J. KORNSTAEDT	Freie Universität Berlin Institut für Mathematik III Arnimallee 2-6, 1000 Berlin 33
W. KRABS	Fachbereich Mathematik, Arbeitsgruppe 10 Technische Hochschule Darmstadt Kantplatz 1, 6100 Darmstadt
K. ROLEFF	Institut für Angewandte Mathematik der Universität, Bundesstr. 55, 2000 Hamburg 13

W.W.E. WETTERLING Technische Hochschule Twente/TW,
Postbus 217, Enschede, Niederlande

P. ZENCKE Institut für Angewandte Mathematik, SFB 72,
Universität Bonn,
Wegelerstr. 6, 5300 Bonn

J. ZOWE Institut für Angewandte Mathematik u. Statistik
Universität Würzburg
Am Hubland, 8700 Würzburg

Introductory remarks on semi-infinite programming

Semi-infinite programming deals with the problem of minimizing a function $f(x_1,...,x_n)$ of variables x_i, which, for every y in some set Y, are required to meet an inequality $g(x_1,...,x_n;y) \leq r(y)$. In case of a finite set Y this is a usual optimization problem with a finite number of inequality constraints on the variables x_i. In many applications infinite sets Y occur quite naturally:

For instance, in controlling air-pollution in some region Y, let x_i be costs incident to provisions for reducing the emission of some substance. Then the total cost $f(x_1,...x_n) = \sum_{i=1}^{n} x_i$ is to be minimized subject to the restriction that in each point $y \in Y$ the concentration $g(x_1,...x_n;y)$ remains below some given standard $r(y)$.

Obviously, there are many problems in numerous applications the constraints of which depend on the time or the space coordinates and therefore may be formulated as semi-infinite problems. Another question is, whether a semi-infinite model should be used in practice instead of a discretized one, the solution of the latter being close to that of the first if sufficiently many points are taken into account. In fact, if it were true that generally a discrete problem could be treated more easily and efficiently, from a practical point of view it would not be worth-while to consider semi-infinite programming. However, there are two aspects which make the problem interesting in practice:

- A model with constraints depending on some parameter y (time, space, etc.) given by means of one inequality $g(x_1,...,x_n;y) \leq r(y)$ rather than a big number of unrelated inequalities $g_i(x_1,...,x_n) \leq r_i$ is preferable with respect to storage requirements as well as simplicity of handling. Even in cases where the second description is given initially (for instance due to measurement restrictions) it may be advantageous to transform the data by some approximation procedure into the first form.

- At the optimal point in general the inequalities are binding (i. e. equality holds) only in a finite number of points $y_1,...,y_r$, with $r \leq n$ as a rule. In many cases, especially if Y is a subset of Euclidian space E^m, $m \geq 2$, this leads to a much higher efficiency of continuous methods, because it is sufficient to consider only a small number of points in each iteration and, moreover, the points for the next step can be computed very efficiently by using derivatives with respect to y for instance.

Together with the great variety of applications (cf. for instance Krabs: Optimierung und Approximation, Teubner, 1975, and the last part of this volume) including problems in fields like

- Chebyshev- and L_1-approximation (including constrained problems)
- Operations Research (cf. the example given above)
- Optimal Control
- Computing bounds for solutions of monotonic boundary value problems
- Free boundary value problems

this justifies a growing interest in semi-infinite programming.

Rainer Hettich

DUALITY THEORY OF SEMI-INFINITE PROGRAMMING

Klaus Glashoff
University of Hamburg
Institute for Applied Mathematics
Bundesstraße 55
D-2000 Hamburg 13
West-Germany

0. Summary and contents

Duality theory of semi-infinite programming (SIP) has its roots in the theory of systems of linear inequalities, in the theory of uniform approximation of functions, and in the classical theory of moments.

In this paper we present the main duality theorems of SIP in such a way that the historical development of this theory is shown and the connections between moment theory and SIP become evident.

Semi-infinite programming is such an important tool in many fields of applications that there is a need for a simple but complete 'standard' way to the main duality theorems - this is what we try to give in our paper.

Our classical 'geometrical' approach to the duality theory of SIP is the natural generalization of the theory of Linear Programming. It is very useful for the understanding and development of numerical methods for SIP (c.f. GLASHOFF and GUSTAFSON (1978), ROLEFF (1979)).

Contents:

1. The primal problem of SIP

Let S be a fixed given nonempty set of indices. Assume that for any s in S there is given a real vector

$$a(s) \in R^n$$

and a real number

$$b(s) \in R.$$

Let $c \in R^n$ be given, too.

Then the primal problem (P) of semi-infinite programming is defined as follows:

$$\text{(P)} \qquad \begin{array}{l} \text{minimize} \quad c^T y \\ \text{subject to the constraints} \\ a(s)^T y \geq b(s) \qquad \text{for all } s \in S. \end{array}$$

In this problem it is required to minimize a linear function of y subject to a (possibly infinite) number of linear inequality constraints.

The set Y of admissible vectors,

$$Y = \{ y \in R^n \mid a(s)^T y \geq b(s) \qquad \text{for all } s \in S \},$$

is easily seen to be convex and closed, because it is the intersection of closed half-spaces.

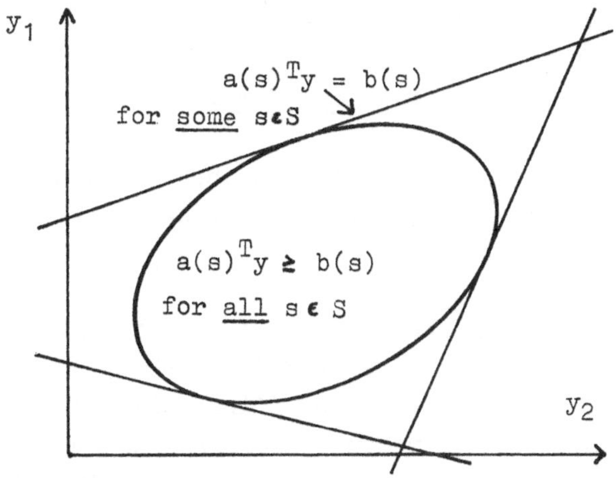

Fig. 1

Problem (P) seems to have a very simple structure but it includes many general optimization problems.

For example, let $f : T \to R$ be an arbitrary real-valued function on an arbitrary set T. Then the problem of finding a global minimum of f on T is trivially equivalent to the following 'simple' SIP:

$$\begin{aligned} &\text{minimize} \quad -y_o \\ &\text{subject to the constraints} \\ &-y_o \geq -f(t) \qquad \text{for all } t \in T. \end{aligned}$$

A nontrivial and by far more practically important example arises if, in the problem given above, $T = R^n$ and $f : R^n \to R$ is convex and (say) differentiable. In this case it is well known that

$$f(y) = \max_{t \in R^n} \{ f(t) + \nabla f(t)^T (y-t) \} ,$$

and minimization of f on R^n is equivalent to the following SIP:

$$\begin{aligned} &\text{minimize } y_o \\ &\text{subject to} \\ &y_o - f(t)^T y \geq f(t) - \nabla f(t)^T t \\ &\qquad\qquad\qquad \text{for all } t \in R^n . \end{aligned}$$

By means of analogeous transformations, it is possible to formulate any convex optimization problem in R^n as a linear SIP.

SIP's are the natural generalization of Linear Programs, where S is a finite set. They are called 'semi-infinite', because there are a finite number of variables but possibly infinitely many constraints.

The first paper explicitely concerning SIP's is that of CHARNES, COOPER and KORTANEK (1962) who use the results of HAAR (1924) on semi-infinite systems of linear inequalities.

In our paper we show that duality theorems of SIP are strongly connected with the theory of moments developed by KREIN (1951), ROSENBLOOM (1952), ROGOSINSKI (1957,1962). In fact, the 'first duality theorem' of SIP appeared in the framework of moment theory and has been independently proved in 1960 by JSII and KARLIN (c.f. KARLIN (1966)).

2. The dual problem

Let $\{ s_1,\ldots,s_q \}$ be a finite subset of S, $q \geq 1$ and $x_1,\ldots,x_q$ nonnegative real numbers such that

$$a(s_1)x_1 + \ldots. + a(s_q)x_q = c .$$

Then, for each vector y which is feasible for (P) ,

$$b(s_1)x_1 + \ldots. + b(s_q)x_q \leq c^T y .$$

This is a very important possibility to construct lower bounds for the value of (P), and it leads us to the dual problem:

$$\text{(D)}\qquad \begin{array}{l} \text{maximize } \sum_{i=1}^{q} b(s_i)x_i \\ \text{subject to the constraints} \\ \sum_{i=1}^{q} a(s_i)x_i = c \\ \left.\begin{array}{l} \{ s_1,\ldots, s_q \} \subset S \\ x_1,\ldots, x_q \geq 0 \end{array}\right\} q \geq 1 . \end{array}$$

One can easily prove that it is possible to fix

$$q = n + 1$$

in the dual problem (apply CARATHEODORY's reduction process to a given n + 1 - vector

$$\sum_{i=1}^{q} \begin{pmatrix} b(s_i) \\ a(s_i) \end{pmatrix} x_i ,$$

where $\{ s_1,\ldots,s_q \}$ and $x_1,\ldots,x_q$ are feasible for (D)).

We shall see that all <u>solutions</u> of (D) can be obtained if we choose

$$q = n .$$

Let us remark that (D) is a <u>nonlinear</u> problem in the variables $s_1,\ldots,s_q$ which can be transformed into a linear problem only if S is a finite set (Linear Programming case).

We denote by v(P) and v(D) the values of (P) and (D), respectively. Then the following 'weak duality'-lemma is a consequence of the considerations discussed above.

<u>Lemma 1:</u> $v(D) \leq v(P)$.

There is another historically important formulation of a dual problem for (P) which shows us the connections of SIP to the theory of moments.

Let us assume for simplicity that S is a closed interval of the real line and that the functions $a_1(s),\ldots,a_n(s)$ are continuous on S. Then we consider the following optimization problem:

(D1)

$$\begin{aligned} &\text{maximize } \int_S b(s)d\alpha(s) \\ &\text{subject to the constraints} \\ &\int_S a_r(s)\, d\alpha(s) = c_r , \quad r = 1,\ldots,n ; \\ &\alpha \text{ is nondecreasing} . \end{aligned}$$

Again it is easily seen that

$$v(D1) \leq v(P) .$$

The precise relation between (D) and (D1) will be discussed in the sequel .

The following theorem is due to ROGOSINSKI (1962) who proved it for the general case in which S is a given set in a finite dimensional real Euclidean space. All integrals are to be understood in the sense of Stieltjes-Lebesgue .

Theorem 1: $$CC(A_S) = M_n \ .$$

An important special case of this theorem was proved by ROSENBLOOM (1952) . ROGOSINSKI derived his interesting results in the framework of KREIN's 'geometrical approach' to the theory of moments (1951) .

In order to apply this theorem to the dual pairs (P)-(D) and (P)-(D_1) , we introduce the following subset of R^{n+1} :

$$\tilde{A}_S = \{ \ (b(s), a_1(s), \ldots, a_n(s))^T \ / \ s \in S \ \} \ .$$

By Theorem 1, the definition of the following moment cone in R^{n+1} will not lead to a confusion:

$$M_{n+1} = CC(\tilde{A}_S) \ .$$

With this notation in mind the dual problems (D) as well as (D_1) can be written in the following 'geometrical' form:

maximize c_o
subject to the constraints
$(c_o, c)^T \in M_{n+1}$.

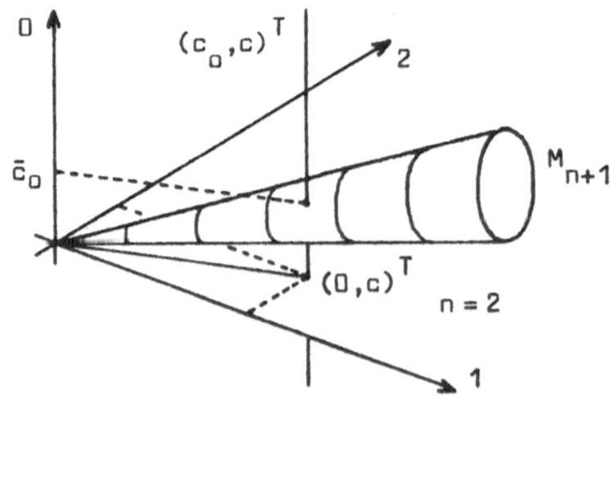

Fig. 2

Geometry of the dual problem

Problem (D1) appeared in a paper by JSII (1960) and in the book of KARLIN (1966) p.470. In the latter, S is a subset of the Euclidean space R^k , and α is a nonnegative finite regular measure defined on S fulfilling the integrability conditions

$$\int_S |a_r(t)|d\alpha(t) \ , \ \int_S |b(t)|d\alpha(t) < \infty$$

for $r = 1,\ldots, n$.

3. Moment cones

The following 'geometrical approach' to the dual problem is due to KREIN (1951) .

Let the moment cone $M_n \subset R^1$ be defined

by

$$M_n = \{ d\varepsilon R^n \ / \ d_r = \int_S a_r(s)d\alpha(s) \ , \ d\alpha \geq 0 \} .$$

Problem (D1) is feasible iff $c \in M_n$.

We obtain a subset of M_n by considering nonnegative linear combinations of point measures. Let

$$A_S = \{ (a_1(s),\ldots,a_n(s))^T \ / \ s \in S \}$$

and let $CC(A_S)$ denote the convex cone generated by A_S :

$$CC(A_S) = \{ \sum_{i=1}^{q} a(s_i)x_i \ / \ s_i \in S \ , \ x_i \geq 0 \} .$$

Then it is easily seen that the following relation holds:

Lemma 2: $CC(A_S) \subset M_n \subset \overline{CC(A_S)}$.

One of the important questions in the general theory of moments concerns the exact relation between $CC(A_S)$ and M_n .

4. The first duality theorem

The following duality theorem has a long history (c.f. KARLIN (1966), p. 467). It belongs to the fundamental results in the field of 'Tchebycheff inequalities' which are of interest and relevance in probability theory, statistical analysis, and elsewhere.

Theorem 2 (ISII, KARLIN 1960): Suppose that

(i) $v(D)$ is finite ;

(ii) $c \in \text{int } M_n$.

Then, $v(D) = v(P)$ (i.e. there is no 'duality gap'), and (P) has a solution .

Remark: (ii) is a 'constraint qualification'.

Proof: The vector $(v(D), c_1 \ldots, c_n)^T$ is a boundary point of M_{n+1}. There exists a supporting hyperplane to M_{n+1} through this vector:

(1) $$y_o v(D) + \sum_{r=1}^{n} y_r c_r = 0 ,$$

(2) $$y_o z_o + \sum_{r=1}^{n} y_r z_r \geq 0 \qquad \text{for all } z \in M_{n+1} .$$

Taking $(z_o, z_1, \ldots, z_n)^T = (b(s), a_1(s), \ldots, a_n(s))^T$, $s \in S$,

in (2), we obtain

$$y_o b(s) + \sum_{r=1}^{n} y_r a_r(s) \geq 0 \qquad \text{for all } s \in S .$$

Now it is easy to prove by means of (1) and (ii) that $y_o < 0$ (KARLIN (1966), p. 473) .

We divide (1) and (2) by $-y_o$ and obtain

$$\sum_{r=1}^{n} \frac{y_r}{y_o} c_r = v(D) ,$$

$$\sum_{r=1}^{n} \frac{y_r}{y_o} a_r(s) \geq b(s) \qquad \text{for all } s \in S .$$

This shows us that the vector $y_o^{-1}(y_1,\dots,y_n)^T$ is optimal for (P) and that there exists no duality gap.

Example: $S = [0,1]$, $a_1(s) = 1$, $a_2(s) = s$,
$b(s) = \sqrt{s}$, $c = (1,0)^T$.

In this example, (i) does not hold; (P) has no solution.

5. The second duality theorem

The second duality theorem is directly implied by the inhomogeneous FARKAS - Lemma (HAAR, 1924) for semi-infinite systems of linear inequalities. We are going to present a 'modern' version and proof of this theorem.

Lemma 3 (HAAR 1924): Suppose that

(i) There is a $\tilde{y} \in R^n$ such that $\Sigma a_r(s)\tilde{y}_r \geq b(s)$ for all $s \in S$.

(ii) M_{n+1} is closed.

Then (a) and (b) below are equivalent:

(a) $\sum_{r=1}^{n} a_r(s)y_r \geq b(s)$ for all $s \in S$ $\Rightarrow$ $c^Ty \geq c_o$

(b) There exist $s_1,\dots,s_{n+1} \in S$, $x_1,\dots,x_{n+1} \geq 0$

such that

$$\sum_{i=1}^{n+1} a(s_i)x_i = c ,$$

$$\sum_{i=1}^{n+1} b(s_i)x_i \geq c_o .$$

Proof: (a) is equivalent to the following statement: There is no solution $(y^T, y_{n+1}, y_{n+2})^T \varepsilon R^{n+2}$ of the system

$$\begin{aligned} a(s)^T y - b(s) y_{n+1} \qquad\qquad &\geq 0 \qquad \text{for all } s \varepsilon S \;, \\ -c^T y + \quad c_o y_{n+1} + y_{n+2} &\geq 0 \\ y_{n+1} + y_{n+2} &\geq 0 \\ y_{n+2} &< 0 \;. \end{aligned}$$

This, in turn, is equivalent to the following assertion: There is no solution $\tilde{y} \varepsilon R^{n+2}$ of the system

$$\begin{aligned} u^T \tilde{y} &\geq 0 \qquad \text{for all } u \varepsilon U \;, \\ v^T \tilde{y} &< 0 \end{aligned}$$

where

$$U := \{ (a(s)^T, -b(s), o)^T \;,\; s \varepsilon S \} \cup \{ (-c^T, c_o, 1)^T \} \cup \{ \theta^T, 1, 1)^T \} \;,$$

$$v = (\theta^T, o, 1)^T \;.$$

By the separating hyperplane theorem ('homogeneous Farkas Lemma'),

(3) $$v \;\varepsilon\; \overline{CC(U)}$$

Now CC(U) is a <u>closed</u> cone, because it is the difference of two closed cones which intersect only at $\theta \;\varepsilon\; R^{n+2}$:

$$CC(U) = K_1 - K_2 \;,$$

where

$$K_1 = CC(\{ (a(s), -b(s), o)^T \;,\; s \;\varepsilon\; S \})$$

is closed by (ii) and

$$K_2 = CC(\{ (c, -c_o, -1)^T \;,\; (\theta, -1, -1)^T \})$$

is closed, too. It is easy to show that $K_1 \cap K_2 = \{ \theta \}$. - As CC(U) is closed, (3) implies

$$v \;\varepsilon\; CC(U) \;,$$

i.e. there exists a subset $\{ s_1,\dots,s_{n+1} \} \subset S$ and nonnegative numbers $x_1,\dots,x_{n+3}$ such that

$$(4) \qquad \begin{pmatrix} \Theta \\ 0 \\ 1 \end{pmatrix} = \sum_{i=1}^{n+2} x_i \begin{pmatrix} a(s_i) \\ -b(s_i) \\ 0 \end{pmatrix} + x_{n+2} \begin{pmatrix} -c \\ c_o \\ 1 \end{pmatrix} + x_{n+3} \begin{pmatrix} \Theta \\ 1 \\ 1 \end{pmatrix} .$$

We show that $x_{n+2} \neq 0$. Otherwise, multiplication of (4) by $(\tilde{y}^T,1,0)^T$ gives

$$0 = \sum_{i=1}^{n+1} x_i(\tilde{y}^T a(s_i) - b(s_i)) + x_{n+3} .$$

All terms are nonnegative, which implies $x_{n+3} = 0$ in contradiction to (4). - Dividing (4) by $x_{n+2} > 0$, we obtain the desired result.

Theorem 3 (Second duality theorem): Suppose that

(i) $v(P)$ is finite ,

(ii) M_{n+1} is closed .

Then, $v(D) = v(P)$, and (D) has a solution .

Proof: This theorem follows directly from the preceding inhomogeneous Farkas-Lemma: Choose $c_o = v(P)$; by (b) , the value $v(D)$ of the dual problem is not smaller than $v(P)$, and the desired result follows by the 'weak' duality Lemma 1 .

In the following section we consider an important condition under which M_{n+1} is closed (it is well known that this is true in case S is a finite set, because then M_{n+1} is finitely generated) .

6. Closed moment cones

As mentioned before, the first paper on the theory of semi-infinite programming was written by Charnes, Cooper, and Kortanek in 1962 (corrected by DUFFIN and KARLOVITZ (1965)) .

In their paper, C.,C., and K. gave a translation of Farkas' paper from 1924 on which they based their duality theorem, the proof of which is not very easy to understand.

In this section we shall prove the C.,C., and K. - theorem by means of the Farkas Lemma and the following result of ROGOSINSKI (1962) which is a special case of a general theory on infinite moment problems.

Lemma 4: Under the following assumptions

(i) S is a compact set; the functions a (•) and b(•) are continuous on S ,

(ii) there exists a $\hat{y} \in R^n$ such that

$$\sum_{r=1}^{n} a_r(s)\hat{y}_r > b(s) \qquad \text{for all } s \in S ,$$

the cone M_{n+1} is closed .

A proof of this result may be found, f.i., in the book of GLASHOFF and GUSTAFSON (1978) .

In the C., C., and K. - theorem, which we are going to prove now, there are no such conditions like (i) in the preceding Lemma - the set S and the functions a(•) and b(•) are quite arbitrary. But we shall see that the assumptions of the following theorem allow us to apply Lemma 4 .

Theorem 4: Assume that

(i) The set of vectors $d(s) = (a(s)^T, b(s))^T \in R^{n+1}$ is "canonically closed"; this means that there exists a set of positive constants k(s) such that the vectors k(s)d(s) , $s \in S$, form a compact set in R^{n+1} ;

(ii) There exists $\hat{y} \in R^n$ such that

$$\sum_{r=1}^{n} a_r(s)\hat{y}_r > b(s) \qquad \text{for all } s \in S .$$

Then, if v(P) is finite, v(P) = v(D), and (D) has a solution.

Proof: The constraints of (P) are equivalent to the following system of inequalities:

$$(5) \qquad k(s)a(s)^T y \geq k(s)b(s) \qquad \text{for all } s \in S.$$

This transformation does not change the value and the solvability of the dual problem . Let M_{n+1} denote the moment cone in R^{n+1} belonging to (5); then

$$\hat{M}_{n+1} = CC(\hat{A}_S) \; .$$

where

$$\hat{A}_S = \{ \; k(s)d(s) \; / \; s \in S \; \}$$

is a compact set. Evidently (5) can be written as follows (again the dual is not 'seriously' changed):

$$t_1 y_1 + t_2 y_2 + \ldots + t_n y_n \geq t_{n+1} \quad \text{for all } (t_1, \ldots, t_{n+1}) \varepsilon \hat{A}_S.$$

The coordinate-functions $t \to t_i$ are trivially continuous, thus we may apply Lemma 4, and the result follows by the second duality theorem.

7. Concluding Remarks

In our paper we have thrown light on an important part of the historical development of SIP, especially pointing out the independent developments and connections of SIP and the theory of moments.

Let us remark that HAAR stated his theorem without assumption (ii) Lemma 3. Charnes, Cooper, and Kortanek showed that Haar implicitely used the assumption of canonical closure of the set of inequalities,and Duffin and Karlovitz made the observation that the 'Slater - condition' (ii), Theorem 4, did not explicitely appear in Farkas' Lemma but was used by C.,C., and K. in the proof of it.

We want to stress that it is of more practical value to treat semi-infinite programming as the natural generalization of Linear Programming than just as a special case of the now well developed theory of infinite optimization (this may be left as an exercise for students in optimization theory). For any theoretical result on SIP can be obtained by applying R^n - theory only - there is absolutely no need for the application of (nonconstructive) separating hyperplane theorems in general locally convex spaces which one uses in infinite optimization .

The relations of SIP to Linear Programming are of special value for the numerical treatment of SIP's - here the Simplex-method is one of the most efficient tools for the practical solution of such problems; c.f. ROLEFF (1979), these Proceedings, and the references given there.

Acknowledgement. I thank Prof.Carl Geiger, University of Hamburg, for interesting discussions on the subject of this paper.

Appendix

In section 2 we remarked that all solutions of (D) are obtained by choosing q = n in the formulation of the dual problem. We are going to give a simple proof for this fact.

Let $\{ \hat{s}_1,\dots,\hat{s}_{n+1} \} \subset S$, $\hat{x}_1,\dots,\hat{x}_{n+1} \geq 0$ be a solution of (D).

Then the vector

$$(v(D),c_1,\dots,c_n)^T = \sum_{i=1}^{n+1} \begin{bmatrix} b(\hat{s}_i) \\ a(\hat{s}_i) \end{bmatrix} \hat{x}_i$$

lies at the boundary of M_{n+1} (c.f. the proof of Theorem 2). But it is very easy to realize that any vector

$$\sum_{i=1}^{n+1} \begin{pmatrix} b(s_i) \\ a(s_i) \end{pmatrix} x_i \ , \qquad x_1,\dots,x_{n+1} > 0 \ ,$$

is in the interior of the moment cone M_{n+1}. This shows us that not all $\hat{x}_i$ are different from zero, which proves the assertion.

References

A.CHARNES, W.W.COOPER, and K.KORTANEK: Duality in semi-infinite programs and some works of Haar and Carathéodory. Management Science, Vol.9, No.2, January 1963, 209-228.

A.CHARNES, W.W.COOPER, and K.KORTANEK: On representation of semi-infinite programs which have no duality gaps. Management Science, Vo. 12, No.1, September 1965, 113-121.

R.J.DUFFIN and L.A.KARLOVITZ: An infinite linear program with a duality gap. Management science, Vol.12, No.1, September 1965, 122-134 .

K.GLASHOFF and S.Å.GUSTAFSON: Einführung in die lineare Optimierung. Wissenschaftliche Buchgesellschaft, Darmstadt, 1978.

S.Å.GUSTAFSON and K.O.KORTANEK: Numerical treatment of a class of semi-infinite programming problems. Naval Research Logistics Quarterly, Vol. 20, No.3, September 1973, 477-504.

A.HAAR: Über lineare Ungleichungen. Acta Math.(Szeged), 2 ,1924 - 1926, 1-14.

K.ISII: The extrema of probability determined by generalized moments (I) Bounded random variables. Ann.Inst.Stat.Math., 12, 1960, 119-133.

S.J.KARLIN and W.J.STUDDEN: Tchebycheff systems: with applications in analysis and statistics. John Wiley & Sons, Inc.,New York 1966.

M.G.KREIN: The ideas of P.L.ČEBYSEV and A.A.MARKOV in the theory of limiting values of integrals and their further developments. Am.Math.Soc.Transl., Ser.2, 12, 1951, 1-122.

W.W.ROGOSINSKI: Moments of non-negative mass.Proc.Roy.Soc. (London), A,245, 1958, 1-27.

W.W.ROGOSINSKI: Non-negative linear functionals, moment problems, and extremum problems in polynomial spaces. In : Studies in mathematical analysis and related topics; G.SZEGÖ (Ed.), Stanford University Press, Stanford, California, 1962.

K.ROLEFF: A stable multiple exchange algorithm for linear SIP. These Proceedings, 1979.

P.C.ROSENBLOOM: Bull. Soc. math. Fr.80,1952, 183-215

Additional references

A.CHARNES, W.W.COOPER, and K.KORTANEK: Duality, Haar Programs, and finite sequence spaces. Proc. National Acad.Sc.USA, 48, No.5, May 1962, 783-786.

U.ECKHARDT: Theorems on the dimension of convex sets. Linear Algebra Appl. 12, 1975, 63-76.

L.L.DINES and N.H.Mc COY: On linear inequalities. Trans.Roy.Soc.Canada 27, 1933, 37-70.

R.J.DUFFIN: An orthogonality theorem of Dines related to moment problems and Linear Programming. J.Combin.Th.2, 1967, 1-26.

W.KRABS: Optimierung und Approximation. Teubner Studienbücher Mathematik, B.G.Teubner, Stuttgart 1975.

SECOND ORDER NECESSARY OPTIMALITY CONDITIONS FOR SEMI-INFINITE PROGRAMMING PROBLEMS

Aharon Ben-Tal
Marc Teboulle
Jochem Zowe*)

Department of Computer Science
Technion-Israel Institute of Technology
Haifa, Israel

ABSTRACT.

A unified second order theory for extremum problems is applied to obtain second order necessary conditions for semi-infinite programming problems. The well known first order conditions are included. An application to the problem of best local nonlinear approximation is given.

1. INTRODUCTION

Consider the Semi-Infinite Programming problem

(S.I.P.): $\inf \; f^o(x)$

subject to $f^k_\alpha(x) \leq 0, \; \alpha \in A_k, \; k \in P,$

where

$P \triangleq \{1,2,\ldots,p\}$;

A_k is a compact subset of $R^\ell (\ell \geq 1)$, $k \in P$;

$f^o: R^n \to R$ is twice continuously differentiable;

$f^k_\alpha: R^n \to R$ is twice differentiable for every $\alpha \in A_k$, $k \in P$; moreover $f^k_\alpha(x)$, $\nabla f^k_\alpha(x)$ and $\nabla^2 f^k_\alpha(x)$ are continuous in x,α where ∇f^k_α and $\nabla^2 f^k_\alpha$ denotes, respectively, the gradient vector and the Hessian of f^k_α.

Many optimization problems fit this formalism, see e.g. the paper by Gustafson and Kortanek, [6].

If we define $f^k(x) = \sup_{\alpha \in A_k} f^k_\alpha(x,\alpha)$, all $k \in P$, then (S.I.P.) becomes a problem with a finite number of constraints:

(P): $\inf \; f^o(x)$

s.t. $f^k(x) \leq 0, \; k \in P.$

The fundamental difficulty, however, with problem (P) is that the constraint functions $f^k(x)$ are typically nondifferentiable. Thus a theory for nondifferentiable nonlinear programming is needed. Such a first order theory does exist (e.g.[5], [10]) and in-

*Permanent address: Institut für Angewandte Mathematik, Universität Würzburg 8700 Wurzburg, W.-Germany.

deed was used to establish first order condition for (S.I.P.); see e.g. [11], [10] and [5]. Here, to obtain second order conditions (involving second derivatives) for (S.I.P.), we use the recent results in [1], whose essentials are introduced in Section 2. These results are used in Section 3 to obtain second order necessary conditions for local optimality in problem (S.I.P.). Section 4 contains an application to best nonlinear local approximation.
A different approach to get second order condition in (S.I.P.) can be found in [7].

2. THE GENERAL SECOND ORDER NECESSARY CONDITIONS

We present here the main features of a second order theory for general extremum problems (restricted to the finite dimensional case and without equality constraints). For further details and proofs see [1].

We begin with some definitions and notations. Let f be a real valued function on R^n. A vector $d \in R^n$ is a direction of quasi-decrease of f at x if, for every $\alpha > 0$, there is a positive scalar T such that:

$$f(x + td) \leq f(x) + \alpha t, \qquad t \in (0,T].$$

The set of all such vectors is denoted by $D_f(x)$. Clearly $D_f(x)$ is a cone. It is easily seen that in terms of the directional derivative,

$$f'(x,d) \triangleq \lim_{t \to 0^+} \frac{f(x+td) - f(x)}{t},$$

this cone is characterized by

Proposition 2.1 If $f'(x,d)$ exists, then

$$d \in D_f(x) \quad \text{if and only if} \quad f'(x,d) \leq 0. \qquad \square$$

Next we define the following "second order" set. For fixed $x,d \in R^n$ let:

$$Q_f(x,d) = \left\{ z \in R^n : \begin{array}{l} \exists\, T > 0, \text{ a neighbourhood } N \text{ of } z \text{ and } \beta > 0 \\ \text{such that for every } \bar{z} \text{ in } N: \\ f(x+td + \frac{1}{2}t^2\bar{z}) \leq f(x) - \beta t^2, \text{ all } t \in (0,T] \end{array} \right\}$$

We put $D_f^<(x) \triangleq Q_f(x,0)$ and note that this is exactly the cone of directions of decrease of f at x as considered e.g. in [5].

A vector which is a direction of quasi-decrease, but is not a direction of decrease, is called a critical direction. The set of all such directions is denoted by $D_f^=(x)$, i.e.:

$$D_f^=(x) = D_f(x) \cap \text{comp}\, D_f^<(x) .$$

A function f is Q-regular at x if:
for every $d \in D_f^=(x)$ the set $Q_f(x,d)$ is convex.

To characterize $Q_f(x,d)$ analytically we define for $x,d,z \in R^n$ (whenever this limit exists)

(2.1) $f''(x,d;z) \triangleq \lim_{t\to 0^+} \frac{1}{t^2}[f(x+td+\frac{1}{2}t^2z)-f(x)]$.

Then, similarly to the characterization of $D_f^<(x)$ by $f'(x,d)$ (see [5], Theorem 7.5) the next result describes $Q_f(x,d)$ in terms of $f''(x,d;z)$; for a proof see [1], Proposition 3.

<u>Proposition 2.2</u> If $z \in Q_f(x,d)$ and the limit (2.1) exists then $f''(x,d;z) < 0$. Conversely, if (2.1) exists and $f''(x,d;z) < 0$ and, if f satisfies at x a local Lipschitzian condition, then $z \in Q_f(x,d)$. □

If f is differentiable then one has the trivial relation: $f'(x,d) = \nabla f(x)d$. The following result shows how $f''(x,d;z)$ is related to $\nabla f(x)$ and $\nabla^2 f(x)$.

<u>Lemma 2.1</u> Let f be twice continuously differentiable. Then $f''(x,d;z)$ exists if and only if $\nabla f(x)d = 0$, in which case

$$f''(x,d;z) = \frac{1}{2}[\nabla f(x)z+d^T\nabla^2 f(x)d].$$

<u>Proof.</u> Let $x, d, z \in R^n$ be given. By the Taylor Theorem, applied to the function $\varphi(s) = f(x+sd+\frac{1}{2}s^2z)$ at $s = 0$, we get for $t > 0$:

$$\begin{aligned} &f(x+td+\tfrac{1}{2}t^2z)-f(x) \\ &\quad = t\nabla f(x)d + \tfrac{1}{2}t^2[\nabla f(x)z+d^T\nabla^2 f(x)d]+o(t^2). \end{aligned}$$

The assertion follows easily from this and (2.1).

The role of the <u>support functional</u> is fundamental in the formulation of our second-order necessary conditions. We recall here only some basic definitions and results; for further details see [1], [12] and [13].

With a given subset S of R^n one associates its so-called <u>indicator function</u> $\delta(\cdot|S):R^n\to R \cup \{\infty\}$,

$$\delta(x|S) = \begin{cases} 0 & \text{if } x \in S \\ \infty & \text{otherwise} \end{cases},$$

and its <u>support function</u> $\delta^*(\cdot|S):R^n \to R\cup\{+\infty, -\infty\}$,

$$\delta^*(y|S) = \sup_{x\in S}\langle x,y\rangle.$$

We will denote the effective domain of $\delta^*(\cdot|S)$ by $\Lambda(S)$, i.e.

$$\Lambda(S) = \{ y \in R^n:\delta^*(y|S)<\infty \}.$$

It is well-known that $\delta^*(\cdot|S)$ is a closed positively homogeneous convex function and $\Lambda(S)$ a closed convex convex cone. Clearly, when S is a convex cone:

$$\Lambda(S) = \{y \in R^n: \langle x,y\rangle \le 0 \text{ all } x \in S\}.$$

The set in the right hand side is the so-called <u>polar cone</u> of S, denoted by S^*.

We are now in a position to formulate the characterization theorem for the general problem

$$(\hat{P}):\quad \inf\ f^{o}(x)$$
$$\text{s.t.} f^{k}(x) \leq 0,\ k \in P=\{1,\ldots p\}$$

where f^k, $k \in \{0\} \cup P$, are continuous real valued functions.

Let us introduce the following notations:

The cone $D_{f^k}(x)$ will be denoted shortly $D_k(x)$; similar abbreviations are used for $\overset{=}{D}_{f^k}$, $Q_{f^k}(x,d)$. Furthermore, define $I(x_o) = \{k \in P: f^k(x_o) = 0\}$ and $I_o(x_o) = \{0\} \cup I(x_o)$. Finally, for $d \in \bigcap_{k\in I_o(x_o)} D_k(x_o)$, let:

$$J(x_o,d) \triangleq \left\{k \in I_o(x_o):\ d \in \overset{=}{D}_k(x_o)\right\}.$$

Theorem 2.1 Let x_o be a local minimum for problem $(\hat{P})$. Assume that the functions f^k, $k \in I_o(x_o)$, are Q-regular at x_o. Then for every $d \in R^n$ such that

(2.2) $\quad d \in D_k(x_o)$ for $k \in I_o(x_o)$,

there exist vectors

(2.3) $\quad y_k \in \Lambda(Q_k(x_o,d))$ for $k \in J(x_o,d)$,

not all zero, which satisfy the Euler-Lagrange equation

(2.4) $$\sum_{k\in J(x_o,d)} y_k = 0$$

and the Legendre inequality

(2.5) $$\sum_{k\in J(x_o,d)} \delta^*(y_k|Q_k(x_o,d)) \leq 0.$$

Remark 2.1 Theorem 2.1 contains the first order Dubovitskii-Milyutin necessary conditions. This can be easily verified by specializing Theorem 2.1 to the case d=0, see [1].

3. SECOND ORDER NECESSARY CONDITIONS FOR PROBLEM (S.I.P.)

As mentioned in the introduction, the semi-infinite programming problem (S.I.P.) is replaced by a problem (P) with a finite number of constraints $f^k(x) \leq 0$, $k \in P$, where $f^k(x) \triangleq \sup_{\alpha\in A_k} f_\alpha^k(x)$. In order to use Theorem 2.1 to obtain second order necessary conditions for problem (P), one must know how to construct, for a functional of the form $f(x) = \sup_{\alpha\in A} f_\alpha(x)$, the following objects:

(i) the cone of directions of quasi-decrease ,

(ii) the set $Q_f(x,d)$,

(iii) the support functional of an infinite intersection of open convex sets.

In accordance with the assumptions made for our problem (S.I.P.) we assume throughout this section:

(3.1) $\begin{cases} A \text{ is a compact subset of } R^{\ell}; \\ f_\alpha \text{ is twice differentiable for every } \alpha \in A \text{ and } f_\alpha(x),\ \nabla f_\alpha(x) \text{ and } \nabla^2 f_\alpha(x) \\ \text{are continuous in } x,\ \alpha. \end{cases}$

For $x,d \in R^n$ we define

(3.2) $$A(x) \triangleq \{\alpha \in A: f(x) = f_\alpha(x)\},$$
$$A(x,d) \triangleq \{\alpha \in A(x): f'(x,d) = f'_\alpha(x,d)\}$$

Clearly,

$$A(x,d) \subset A(x) \subset A.$$

Furthermore, (3.1) implies that $A(x)$ and $A(x,d)$ are non-empty and compact sets (see, e.g., [11]); we will make use of this in the sequel without further mentioning.

For $\varepsilon > 0$ let $A_\varepsilon(x,d)$ denote an ε-neighbourhood of $A(x,d)$:

$$A_\varepsilon(x,d) = \{\alpha \in A : \min_{\bar{\alpha} \in A(x,d)} ||\alpha - \bar{\alpha}|| \le \varepsilon\} .$$

The next result says that $A(x,d)$ depends 'continuously' on x.

Lemma 3.1 Let $x_o, d \in R^n$ be given. Then for every $\varepsilon > 0$ there exists $\delta(\varepsilon) > 0$ such that $A(x,d) \subset A_\varepsilon(x_o,d)$ whenever $||x - x_o|| \le \delta(\varepsilon)$.

Proof. Suppose the assertion does not hold. Then for some $\varepsilon > 0$ and for all $\delta = \frac{1}{n}$, $n \in N$, there exist points $x_n \in R^n$ and $\alpha_n \in A$ such that $||x_n - x_o|| \le \delta = \frac{1}{n}$, $\alpha_n \in A(x_n,d)$ but $\alpha_n \notin A_\varepsilon(x_o,d)$, i.e.

(3.3) $$\min_{\bar{\alpha} \in A(x_o,d)} ||\alpha_n - \bar{\alpha}|| > \varepsilon \quad \text{for } n \in N.$$

Since A is compact we can assume that $\{\alpha_n\}$ (or more precisely: a suitable subsequence) converges to some $\alpha_o \in A$. Now, by definition of $A(x_n,d)$,

$$f'_{\alpha_n}(x_n,d) = f'(x_n,d).$$

By assumption, $f'_\alpha(x,d) = \nabla f_\alpha(x) d$ is continuous in x, α. Hence, passing to the limit, we get

$$f'_{\alpha_o}(x_o,d) = f'(x_o,d),$$

i.e. $\alpha_o \in A(x_o,d)$. Together with (3.3) we see

$$||\alpha_n - \alpha_o|| > \varepsilon \quad \text{for all } n \in N,$$

contradicting $\alpha_n \to \alpha_o$. □

Our next result shows how to compute $f''(x_o,d;z)$ from $f''_\alpha(x_o,d;z)$. To this end the following assumption is made

P(d): There exists a neighbourhood N of d and $T > 0$ such that $\nabla f_\alpha(x_o) d \le 0$ for all $\alpha \in A(x_o + t\bar{d})$ where $t \in [0,T]$ and $\bar{d} \in N$.

Lemma 3.2 Let $d \in D_f^=(x_o)$ and suppose d satisfies P(d). Then for all z

$$f''(x_o,d;z) = \max_{\alpha\in A(x_o,d)} f''_\alpha(x_o,d;z).$$

Proof. For $d \in D_f^=(x_o)$ we have $f'(x_o,d)= 0$ which implies $\nabla f_\alpha(x_o)d = f'_\alpha(x_o,d) = 0$ for $\alpha \in A(x_o,d)$. Hence by Lemma 2.1, $f''_\alpha(x_o,d;z)$ exists and

$$f''_\alpha(x_o,d;z) = \frac{1}{2}(\nabla f_\alpha(x_o)z + d^T\nabla^2 f_\alpha(x_o)d),\ \alpha \in A(x_o,d)\ . \tag{3.4}$$

Now put $x_t \triangleq x_o+td + \frac{1}{2}t^2z$ for $t > 0$. Then by definition of $A(x_t)$,

$$f_{\alpha_t}(x_t) = f(x_t) \ge f_{\alpha_o}(x_t) \quad \text{for } \alpha_t\in A(x_t),\ t\ge 0,$$

$$f_{\alpha_t}(x_o) \le f(x_o) = f_{\alpha_o}(x_o) \quad \text{for " " "}.$$

Since $A(x) \supset A(x,d)$ the above inequalities remain true with $A(x_t)$ replaced by $A(x_t,d)$. Subtraction of the two inequalities and division by t^2 gives for $\alpha_t \in A(x_t,d)$, $t > 0$, and $\alpha_o \in A(x_o,d)$:

$$\frac{1}{t^2}[f_{\alpha_t}(x_t)-f_{\alpha_t}(x_o)] \ge \frac{1}{t^2}[f(x_t)-f(x_o)] \ge \frac{1}{t^2}[f_{\alpha_o}(x_t)-f_{\alpha_o}(x_o)]. \tag{3.5}$$

The right-hand side inequality implies

$$f''_{\alpha_o}(x_o,d;z) \le \underline{\lim}_{t\to 0^+} \frac{1}{t^2}[f(x_t)-f(x_o)] \quad \text{for all } \alpha_o\in A(x_o,d). \tag{3.6}$$

By our general assumption, $f_\alpha(x)$ is twice continuously differentiable in x. Hence we get from Taylor's Theorem for some $0\le \theta_t\le 1$:

$$\begin{aligned} &f_{\alpha_t}(x_t) - f_{\alpha_t}(x_o) = \\ &= \nabla f_{\alpha_t}(x_o)(td+ \tfrac{1}{2}t^2z) + \tfrac{1}{2}(td+ \tfrac{1}{2}t^2z)^T\nabla^2 f_{\alpha_t}(x_o+\theta_t(x_t-x_o))(td+\tfrac{1}{2}t^2z). \end{aligned} \tag{3.7}$$

Now for $t> 0$ sufficiently small, $x_t = x_o+t\bar{d}$ with a suitable $\bar{d} \in N$ and thus because of P(d), $\nabla f_{\alpha_t}(x_o)d \le 0$. Hence we get from the left-hand side inequality of (3.5) and from (3.7) for $\alpha_t \in A(x_t,d)$, $t > 0$ sufficiently small:

$$\frac{1}{2}[\nabla f_{\alpha_t}(x_o)z+d^T\nabla^2 f_{\alpha_t}(x_o+\theta_t(x_t-x_o))d] + O(t) \ge \frac{1}{t^2}[f(x_t)-f(x_o)]. \tag{3.8}$$

From Lemma 3.1 it follows easily that, for a suitable sequence of t-values converging to 0, the α_t tend to some $\alpha \in A(x_o,d)$. Furthermore, every cluster point of $\{\alpha_t\}$ as $t \to 0$ belongs to $A(x_o,d)$. Hence (3.6) together with Lemma 2.1 and our general assumption (3.1) implies

$$\sup_{\alpha\in A(x_o,d)} f''_\alpha(x_o,d;z) = \sup_{\alpha\in A(x_o,d)} \frac{1}{2}(\nabla f_\alpha(x_o)z+d^T\nabla^2 f_\alpha(x_o)d) \ge \overline{\lim}_{t\to 0^+}\frac{1}{t^2}[f(x_t)-f(x_o)]\ ,$$

Combining this with (3.6) we get the desired result. □

Remark. A closer inspection of the above proof shows that, without an additional assumption on d like P(d), one gets only the inequality

$$f''(x_o,d;z) \geq \max_{\alpha\in A(x_o,d)} f''_\alpha(x_o,d;z) .$$

Examples show that this inequality can be strict. Put, e.g.,

$$f_\alpha(x) \triangleq -(x-\alpha)^2 \quad \text{for} \quad \alpha \in [-1,1],\ x \in R .$$

An easy calculation shows that for $-1 < x < 1$ and $d \neq 0$, P(d) is not satisfied and that for these x and d and arbitrary z

$$f''(x,d;z) = 0 > -d^2 = \max_{\alpha\in A(x,d)} f''_\alpha(x,d;z) .$$

With the help of Lemma 3.2 we can now characterize the sets $Q_f(x,d)$.

Proposition 3.1 Let $d \in D^=_f(x_o)$ and suppose P(d) holds for d. Then

$$Q_f(x_o,d) = \bigcap_{\alpha\in A(x_o,d)} \{z: \nabla f_\alpha(x_o)z < -d^T\nabla^2 f_\alpha(x_o)d\} .$$

Proof. By Lemma 2.1 and Lemma 3.2

$$f''(x_o,d;z) = \max_{\alpha\in A(x_o,d)} \frac{1}{2}(\nabla f_\alpha(x_o)z + d^T\nabla^2 f_\alpha(x_o)d).$$

As f satisfies at x_o a local Lipschitzian condition (see e.g.[5]. Example 7.5), the assertion follows from Proposition 2.2.

The next result will be concerned with the support functional of an infinite intersection of open halfspaces. Before, we need the following result, which follows easily from Lemma 5.8 in [5].

Lemma 3.3 Let $\{K_\alpha\}_{\alpha\in A}$ be a family of open convex cones such that $\bigcap_{\alpha\in A} K_\alpha \neq 0$. Then

$$(\bigcap_{\alpha\in A} K_\alpha)^* = \text{cl conv} (\bigcup_{\alpha\in A} K^*_\alpha)$$

Proposition 3.2 Let $\alpha \to (a_\alpha,\beta_\alpha)$ be a continuous map from the compact set A into $R^n \times R$. Put

$$S_\alpha \triangleq \{x \in R^n: \langle a_\alpha, x\rangle < \beta_\alpha\}$$

and suppose that $\bigcap_{\alpha\in A} S_\alpha$ is nonempty. Then for $y \in \Lambda(\bigcap_{\alpha\in A} S_\alpha)$ there are $\alpha_1,\dots,\alpha_{n+1}$ in A and nonnegative scalars $\eta_1,\dots,\eta_{n+1}$ such that

$$y = \sum_{i=1}^{n+1} \eta_i a_{\alpha_i} \quad \text{and} \quad \delta^*(y \mid \bigcap_{\alpha\in A} S_\alpha) = \sum_{i=1}^{n+1} \eta_i \beta_{\alpha_i}$$

Proof. For $\alpha \in A$ we define the open convex cones

$$K_\alpha \triangleq \{\lambda(s,-1): s \in S_\alpha, \lambda > 0\} .$$

Then, as is easily seen,

$$K_\alpha^* = \{\gamma(a_\alpha,\beta_\alpha) + (0,\varepsilon) : \gamma \ge 0,\ \varepsilon \ge 0\} .$$

Furthermore,

$$y \in \Lambda(\bigcap_{\alpha\in A} S_\alpha) \text{ and } \delta^*(y \mid \bigcap_{\alpha\in A} S_\alpha) \le \mu$$

(3.9) if and only if

$$(y,\mu) \in \{\lambda(s,-1) : s \in \bigcap_{\alpha\in A} S_\alpha,\ \lambda > 0\}^* = (\bigcap_{\alpha\in A} K_\alpha)^* .$$

By assumption, $\bigcap_{\alpha\in A} K_\alpha \ne \emptyset$, and thus by Lemma 3.3,

$$(\bigcap_{\alpha\in A} K_\alpha)^* = \text{cl conv } (\bigcup_{\alpha\in A} K_\alpha^*) . \tag{3.10}$$

We want to show that the closure operation in (3.10) can be omitted. To see this note that

$$\text{conv } (\bigcup_{\alpha\in A} K_\alpha^*) = \left\{ \sum_{i=1}^m \lambda_i\gamma_i(a_{\alpha_i},\beta_{\alpha_i}) + (0,\varepsilon) : \begin{array}{l} m \in N,\ \lambda_i \ge 0,\ \Sigma\lambda_i = 1 \\ \gamma_i \ge 0,\ \alpha_i \in A,\ \ \varepsilon \ge 0 \end{array} \right\} \tag{3.11}$$

$$= \text{cone } B + \text{cone } (0,1),$$

where $B \triangleq \text{conv } \{(a_\alpha,\beta_\alpha) : \alpha \in A\}$. Now, $\{(a_\alpha,\beta_\alpha) : \alpha \in A\}$ is a compact set as the image of A under the continuous map $\alpha \to (a_\alpha,\beta_\alpha)$. But then also B is compact and it follows from a well-known theorem (see e.g. [9], IV Theorem 2.1) that cone B + cone (0,1) is closed if we can show

$$\text{cone } B \cap \text{cone } (0,-1) = \{(0,0)\}. \tag{3.12}$$

To see this, assume

$$\gamma \sum_{i=1}^m \lambda_i a_{\alpha_i} = 0 \quad \text{where} \quad \gamma > 0,\ \lambda_i \ge 0,\ \sum_{i=1}^m \lambda_i = 1.$$

By assumption, there exists $\bar{x}$ with $\langle a_\alpha,\bar{x}\rangle < \beta_\alpha$ for all $\alpha \in A$. It follows

$$\gamma \sum_{i=1}^m \lambda_i \beta_{\alpha_i} > \gamma \sum_{i=1}^m \lambda_i \langle a_{\alpha_i},\bar{x}\rangle = 0,$$

which proves (3.12). Hence, for our special sets K_α the relation (3.10) reduces to

$$(\bigcap_{\alpha\in A} K_\alpha)^* = \text{conv } (\bigcup_{\alpha\in A} K_\alpha^*) .$$

Furthermore, the Theorem of Caratheodory says that one can always choose $m = n+1$ in (3.11). This, together with (3.9) and (3.11) proves the assertion.

□

Before stating our main theorem recall that, by definition

$$f^k(x) = \sup_{\alpha\in A} f^k_\alpha(x) \quad \text{for} \quad k \in P,$$

$I(x) = \{k \in P: f^k(x) = 0\}$, $I_o(x) = \{0\} \cup I(x)$ and for $d \in \bigcap_{k\in I_o(x)} D_k(x)$

$$J(x,d) = \{k \in I_o(x): d \in D^{=}_k(x)\} .$$

Define $A_k(x)$ and $A_k(x,d)$ for $k \in P$ quite analogously to $A(x)$, $A(x,d)$ in (3.1) and put $A_o(x) = A_o(x,d) = \{0\}$. Moreover, replace in P(d) the function f_α by f^k_α and A by A_k.

Theorem 3.1 Let x_o be a local minimum of Problem (S.I.P). Let $d \in R^n$ be such that

(3.13) $$\nabla f^k_\alpha(x_o)d < 0 \quad \text{for} \quad \alpha \in A_k(x_o) \quad \text{and} \quad k \in I_o(x_o)$$

and suppose P(d) holds for $k \in J(x_o,d)$. Then there is a subset Ω of $\bigcup_{k\in J(x_o,d)} A_k(x_o,d)$ containing n+1 indices (not necessarily distinct) and correspondingly n+1 nonnegative multipliers λ_α, $\alpha \in \Omega$, not all zero, such that

(3.14) $$\sum_{k\in J(k_o,d)} \sum_{\alpha\in\Omega\cap A_k(x_o,d)} \lambda_\alpha \nabla f^k_\alpha(x_o) = 0$$

and

(3.15) $$d^T\{ \sum_{k\in J(x_o,d)} \sum_{\alpha\in\Omega\cap A_k(x_o,d)} \lambda_\alpha \nabla^2 f^k_\alpha(x_o)\}d \geq 0.$$

Proof. Consider problem (P) or $(\hat{P})$, respectively, associated with (S.I.P.). Now, $(f^k)'(x_o,d) = \sup_{\alpha\in A_k(x_o)} (f^k_\alpha)'(x_o,d)$ (see [11]). Hence (3.13) implies (2.2) of Theorem 2.1 and thus 2.1 holds with some y_k, $k \in J(x_o,d)$ - (the Q-regularity of f^k, $k \in I_o(x_o)$, is used in the proof of Theorem 2.1 only to guarantee the convexity of the sets Q_k, $k \in J(x_o,d)$; see [1]. In our situation the convexity of these sets follows immediately from Prop. 3.1). For $k \in J(x_o,d)$ we get from Proposition 3.1

$$Q_k(x_o,d) = \bigcap_{\alpha\in A_k(x_o,d)} \{z \in R^n: \nabla f^k(x_o)z < -d^T\nabla^2 f^k(x_o)d\}.$$

Now, Proposition 3.2 says that there are indices $\alpha_{k_i} \in A_k(x_o,d)$, i=1,...,n+1, and correspondingly nonnegative scalars $_{k_1},\ldots, _{k_{n+1}}$ such that

$$y_k = \sum_{i=1}^{n+1} \eta_{k_i} \nabla f^k_{\alpha_{k_i}}(x_o),$$

$$\delta^*(y_k|Q_k(x_o,d)) = - \sum_{i=1}^{n+1} \eta_{k_i} d^T \nabla^2 f^k_{\alpha_{k_i}}(x_o)d.$$

Summing up, (2.4) and (2.5) yield

$$\sum_{k\in J(x_o,d)} \sum_{i=1}^{n+1} \eta_{k_i} \nabla f^k_{\alpha_{k_i}}(x_o) = 0 \quad , \tag{3.16}$$

$$d^T\{ \sum_{k\in J(x_o,d)} \sum_{i=1}^{n+1} \eta_{k_i} \nabla^2 f^k_{\alpha_{k_i}}(x_o)\}d \geq 0. \tag{3.17}$$

In terms of the (affine-) linear functionals

$$g_{k_i}(x) \triangleq \nabla f^k_{\alpha_{k_i}}(x_o)x + d^T \nabla^2 f^k_{\alpha_{k_i}}(x_o)d, \ x\in R^n,$$

(3.16), (3.17) imply

$$\sum_{k\in J(x_o,d)} \sum_{i=1}^{n+1} \eta_{k_i} g_{k_i}(x) \geq 0 \quad \text{for al} \quad x \in R^n \tag{3.18}$$

Note that in (3.18) not all η_{k_i} can be zero (otherwise $y_k = 0$ for all k). By a well-known Alternative Theorem (see e.g. [13] Theorem (6.10.1)) (3.18) is equivalent to the inconsistency of the system

$$g_{k_i}(x) < 0, \quad k \in J(x_o,d), i=1,\dots,n+1. \tag{3.19}$$

Now apply Helly's Theorem (see [13] Theorem (3.7.1)) to see that (3.19) remains unsolvable if the set of indices is restricted to a suitable finite subset, say Ω, consisting of $n+1$ elements only; in other words:

$$g_\alpha(x) < 0 \ , \ \alpha \in \Omega, \tag{3.20}$$

is inconsistent as well. We apply once more the above mentioned Theorem of the Alternative to see that with suitable nonnegative $\lambda_\alpha, \alpha\in\Omega$, not all zero

$$\sum_{\alpha\in\Omega} \lambda_\alpha g_\alpha(x) \geq 0 \quad \text{for all} \quad x \in R^n$$

Going back to the definition of $g_{k_i}(x)$ it is easy to see that the last inequality implies (3.14) and (3.15).

□

As mentioned in the introduction, by substituting $d = 0$ in Theorem 3.1, we obtain the well-known first order necessary condition for (S.I.P.) as given in [3], [8], [11].

Theorem 3.2 Let x_o be a local minimum for (S.I.P.). Then there is a finite subset of $\bigcup_{k\in I_o(x_o)} A_k(x_o)$ of n+1 elements (not necessarily distinct) and correspondingly n+1 nonnegative multipliers $\lambda_\alpha, \alpha\in\Omega$, not all zero, such that

$$(3.21) \qquad \sum_{k\in I_o(x_o)} \sum_{\alpha\in\Omega\cap A_k(x_o)} \lambda_\alpha \nabla f_\alpha^k(x_o) = 0$$

Proof. Obviously, (3.13) and P(d) hold for d=0. Since $A_k(x_o,0) = A_k(x_o)$ and $J(x_o,0) = I_o(x_o)$ the relation (3.14) reduces to (3.21). □

Remark. If the index sets A_k are finite then P(d) is superfluous since the set of d's satisfying (3.13) contains the set of d's satisfying P(d). This follows, since for an index set A of isolated points, A(x) contains A(x+td) for t > 0 small enough.

It may happen that for every vector d, considered in Theorem 3.1 the multiplier of the objective function $\lambda_o \triangleq \lambda_o(d)$ is equal to zero. In this case the necessary conditions are trivial in the sense that they do not depend at all on the objective function. To assure $\lambda_o(d) \neq 0$ we need an additional assumption on the constraints. The following condition CQ(d) is an example for such a constraint qualification

CQ(d): There is $\bar{x}$ such that $\nabla f_\alpha^k(x_o)\bar{x} < 0$ for all $\alpha \in A_k$, $0 \neq k \in J(x_o,d)$.

For d=0 this can be considered as a semi-infinite version of the so-called Arrow-Hurwicz-Uzawa constraint qualification, and in the convex case it coincides with Slater's condition (see, e.g. [3]).

Proposition 3.3 If for the element d considered in Theorem 3.1 the condition CQ(d) is satisfied then $\lambda_o(d) \neq 0$.

Proof. Obviously, $0 \in \Omega$ since otherwise the system (3.20) in the proof of Theorem 3.1 would be consistent because of CQ(d). Now, assume $\lambda_o(d)=0$. Then CQ(d) together with (3.14) implies $\lambda_\alpha=0$ for all $\alpha \in \Omega$ which contradicts the statement of Theorem 3.1.

4. AN APPLICATION TO THE PROBLEM OF BEST LOCAL NONLINEAR APPROXIMATION

There is a strong connection between approximation problems and optimization problems, see e.g. [4]. Several papers have been written on the characterization of solutions of special approximations problems, with particular type of constraints (or with none) via optimization theory, e.g. [4], [10], [11].

Consider the following general nonlinear Tchebycheff approximation problem:

(A) $\begin{cases} \text{Find a vector } x \in R^n \text{ such that } \max_{t \in T} |f(t) - \phi(x,t)| \text{ is minimized, where} \\ T \subset R^\ell \text{ is a compact set, } f(t) \text{ a continuous real valued function and} \\ \phi(\cdot,t) \text{ is twice continuously differentiable in } x \text{ for all } t \in T. \end{cases}$

This can be rewritten equivalently as an optimization problem with a continuum of constraints:

$(\hat{A})$:

$$\min \; x_{n+1}$$

$$\left.\begin{array}{l} f(t) - \phi(x,t) - x_{n+1} \leq 0 \\ -f(t) + \phi(x,t) - x_{n+1} \leq 0 \end{array}\right\} \quad \text{all } t \in T$$

$$x \in R^n.$$

A vector x^o which solves problem (A), and so $(\hat{A})$, will be called a <u>best local approximation</u>.

Of course one is only interested in approximation problems for which $f(t)$ itself cannot be expressed as $f(t) = \phi(x,t)$ for some x. This implies that in problem $(\hat{A})$, $x^o_{n+1} > 0$. The latter means that in $(\hat{A})$ only one of the two constraints can be active for a given $t \in T$.

The results of Section 3 are applicable for problem (A), and can produce second order necessary conditions for a local best approximation (as well as first order condition). Before stating the result we need some notations:

$$T(x^o) \overset{\Delta}{=} t \in T: |f(t) - \phi(x^o,t)| = \max_{t \in T} |f(t) - \phi(x^o,t)|$$

$$T(x^o,d) \overset{\Delta}{=} \{t \in T(x^o): |\nabla\phi(x^o)d| = \max_{t \in T(x^o)} |\nabla\phi(x^o,t)d|\}$$

$$\theta_t = \text{sign}(\phi(x^o,t) - f(t)), \quad \theta_i \overset{\Delta}{=} \theta_{t_i}.$$

<u>Theorem 4.1</u> Let $x^o \in R^n$ be a local best approximation for problem (A). Then for every d for which there exist a neighbourhood N(d) and a positive scalar $\bar{\eta}$, so that

$$\theta_t \nabla\phi(x^o,t)d \leq 0 \quad \text{for all} \quad \begin{cases} t \in T(x_o + \eta\bar{d}) \\ \eta \in [0,\bar{\eta}] \\ \bar{d} \in N \end{cases}, \tag{4.1}$$

there are n+1 points $t_i \in T(x^o,d)$, $i=1,\ldots,n+1$ (not necessarily distinct) and correspondingly n+1 multipliers λ_i, not all zero, such that

$$\sum_{i=1}^{n+1} \theta_i \lambda_i \nabla\phi(x^o,t_i) = 0 \tag{4.2}$$

$$d^T\{\sum_{i=1}^{n+1} \theta_i \lambda_i \nabla^2\phi(x^o,t_i)\}d \geq 0. \tag{4.3}$$

Proof. We prove the assertion by applying the necessary conditions of Theorem 3.1 to problem $(\hat{A})$. The direction vectors in R^{n+1} will be written as (d,d_{n+1}) , $d \in R^n$, $d_{n+1} \in R$.

If the first constraint in $(\hat{A})$ is active $(\theta_t = -1)$, (3.13) becomes

$$d_{n+1} \leq 0$$

$$-\ \phi(x^o,t)d \leq d_{n+1}$$

while, when the second constraint is active $(\theta_t=1)$

$$d_{n+1} \leq 0$$

$$\phi(x^o,t)d \leq d_{n+1} .$$

Together, this can be written as

$$\theta_t \nabla\phi(x^o,t)d \leq d_{n+1} \leq 0 . \tag{4.4}$$

Now, the variable x_{n+1} appears in $(\hat{A})$ only linearly. So, d_{n+1} will not affect the second order condition and one can put $d_{n+1} = 0$. Hence (3.13) becomes for problem $(\hat{A})$

$$\theta_t\nabla \phi(x^o,t)d \leq 0 .$$

Comparing this with (4.1) we see that (4.1) is nothing else but (3.13) and condition P(d) put together.

Now, by Theorem 3.1 there exist (n+2) points $t_i \in T(x^o,d)$ and multipliers λ_i Note that, since $d_{n+1} = 0$, which is the directional derivative of the objective function of $(\hat{A})$ one has $0 \in J(x^o,d)$ and so (3.14) here becomes

$$\lambda_o\begin{pmatrix}\vec{0}\\1\end{pmatrix} + \sum_{i=1}^{n+1} \lambda_i\begin{pmatrix}\theta_i\nabla\phi(x^o,t_i)\\-1\end{pmatrix} = \begin{pmatrix}\vec{0}\\0\end{pmatrix} . \tag{4.5}$$

The last row of (4.5) is:

$$\lambda_o = \sum_{i=1}^{n+1} \lambda_i .$$

Now λ_o, λ_i are non-negative, not all zero, and consequently $\lambda_o > 0$. (Note that, in fact, condition CQ(d) of Section 3 holds here trivially). Finally, since $\nabla^2 f^o(x_o)$ is the zero matrix (here $f^o(x) = x_{n+1}$) then (3.15) is equivalent to (4.3) and this completes the proof.

By substituting d = 0, in Theorem 4.1 we obtain a first order necessary condition.

Theorem 4.2 Let x^o be a local best approximation for problem (A). Then there exist n+1 points $t_i \in T(x^o)$, i=1,...,n+1 and correspondingly n+1 multipliers $\lambda_i \geq 0$ not all zero such that

$$\sum_{i=1}^{n+1} \theta_i \lambda_i \nabla\phi(x^o,t_i) = 0 \qquad \square$$

Clearly, Theorem 4.2 includes the results on first order conditions in linear approximation problems, i.e. when $\phi(x,t) \overset{\Delta}{=} \sum_{i=1}^{n} x_i g_i(t)$, see e.g. [2], [11].

REFERENCES

[1] Ben-Tal, A., "Second order theory of extremum problems", Proceedings of the International Symposium on External Methods and System Analysis (Fiacco, A.V., Kortanek, K., Editors), University of Texas, Austin, 1977, (to appear).

[2] Cheney, E., Introduction to approximation theory, McGraw Hill, New York, 1966.

[3] Gehner, K.R., "Necessary and sufficient optimality conditions for the Fritz-John problem with linear equality constraints", SIAM J.Control,12,pp.140-149.

[4] Gehner, K.R., "Characterization theorems for constrained approximation problems via optimization theory", J.Approximation Theory, 14, pp.51-76 (1975).

[5] Girsanov, I., Lectures on the Mathematical Theory of Extremum Problems, Lecture notes in Econ. and Math.Systems, #67 Springer Verlag, New-York, 1972.

[6] Gustafson, S.A., Kortanek, K.O., "Numerical treatment of a class of semi-infinite programming problems",Nav.Res.log.Quar., 20(1973),pp.477-504).

[7] Hettich, R.T., Jongen, A.Th., "Semi-infinite programming: Conditions of optimality and applications", Optimization Techniques (Stoer, J., Editor), Lecture Notes in Control and Information Sciences, Springer, 1978.

[8] John, F., "Extremum problems with inequalities as subsidiary conditions", Studies and Essays, Courant Anniversary Volume, Interscience, New York, 1968, pp, 187-204.

[9] Krabs, W., Optimierung und Approximation, Teubner, Stuttgart, 1975.

[10] Laurent, J.P., Optimisation et Approximation, Herman, Paris, 1972

[11] Pchenichnyi, B., Necessary conditions for an Extremum, Marcel Dekker, New York, 1970.

[12] Rockafellar, R.T., Convex Analysis, Princeton University Press, Princeton, 1970.

[13] Stoer, J., Witzgall, C., Convexity and Optimization in finite dimensions I., Springer, 1970.

NECESSARY CONDITIONS OF HIGHER ORDER FOR SEMI-INFINITE PROGRAMMING

Hans-Joachim Kornstaedt

In order to obtain necessary optimality conditions of higher order in terms of various types of differentials we extend the abstract variational theory of order m given in [8].

We first formulate a very general optimization problem which contains semi-infinite programming problems as a special case. In the second section we introduce the essentials of the calculus of variations in topological vector spaces. The advantage of our concept of variational sets and variational derivatives is that it leads to a simple development of necessary optimality conditions of order m. In particular, an invariance property shows that the concept of variational derivatives is well adapted to the notion of variational sets. Moreover, for the variational derivatives the chain rule holds without additional assumptions. In the last section we consider the maximum function which plays a key role in many applications, especially in semi-infinite programming problems. We determine various conditions for which the maximum function is differentiable in our sense.

1. Infinite Dimensional Optimization Problems in Linear Spaces.

Throughout this section let X and Y be real linear spaces, let $A \subset B \subset X$ and $Q \subset Y$ be nonempty subsets, and let $f: B \to Y$ be a mapping.

We consider the following nonlinear optimization problem. Minimize f(x) subject to $x_0 \in A$ in the following sense.

Definition. An element $x_0 \in X$ is called a Q-*minimal element of* A *with respect to* f, if

(1) $$x_0 \in A,\ Q \cap (f(A) - f(x_0)) = \phi \ .$$

This terminology is justified by the following examples.

E 1. In case Y is the real line $\mathbb{R}$ and $Q := \mathbb{R}^- := \{y \in \mathbb{R} | y < 0\}$ we have a *scalar minization problem*. $x_0 \in A$ is a Q-minimal element of A with respect to the functional f if and only if $f(x_0) \leq f(x)$ for all $x \in A$.

E 2. Let $(Y, \leq)$ be a partially ordered linear space, and let $P := \{y \in Y | y \leq \theta\}$, $P \cap (-P) = \{\theta\}$ the convex cone which induces the given vector ordering. If $Q := P \setminus \{\theta\}$ then $x_0 \in A$ is a Q-minimal element of A with respect to f if and only if x_0 is *Pareto-minimal*.

E 3. Let Z be another real linear space, let $K \subset Z$ be a nonempty subset and let $\varphi : B \to \mathbb{R}$ and $\psi : B \to Z$ be mappings. We consider the *scalar optimization problem with explicit constraints*

(2) $$\underset{x}{\text{minimize}} \{\varphi(x) \mid x \in A,\ \psi(x) \in K\}.$$

$x_0 \in X$ is a solution of the program (2) if and only if x_0 satisfies the relation (1), where

$$f := (\varphi,\psi) : B \to \mathbb{R} \times Z =: Y \text{ and } Q := \mathbb{R}^- \times (K - \psi(x_0)).$$

E 4. Let S be a nonempty set and let $\varphi : B \to \mathbb{R}$ and $\Psi : B \times S \to \mathbb{R}$ be mappings. The scalar optimization problem

(3) $$\underset{x}{\text{minimize}} \{\varphi(x) \mid x \in A,\quad \forall s \in S\ \ \Psi(x,s) \leq 0\}$$

is called *semi-infinite* if X is finite dimensional.

If the maximum function $\psi : B \to \mathbb{R}$ defined by $\psi(x) := \max_{s \in S} \Psi(x,s)$ exists, then problem (3) is equivalent to program (2), where $Z := \mathbb{R}$ and $K := \mathbb{R}_0^-$.

Another typical example for semi-infinite programming is the wellknown T-approximation problem, which we will consider in section 3 as an important special case.

Remark. The relation (1) suggests the application of (algebraic) separation theorems. Under certain hypotheses one can obtain a global necessary condition of the following form.

There exists a "monotone" linear functional $l \in Y'$ (the algebraic dual of Y), $l \neq \theta$, such that

(4) $$\langle q,l\rangle \leq 0 \leq \langle f(x) - f(x_0),l\rangle \quad , \quad q \in Q,\ x \in A.$$

The relation (4) is called *minimum (maximum) principle* for the problem (1).

2. Local Necessary Conditions.

In developing local necessary conditions for the problem (1) we follow the basic ideas of the calculus of variations. Let us consider the values of a functional $f : B \to \mathbb{R}$ along certain curves $a : (0,\eta) \to A$, $\eta > 0$, such that

$$a(\lambda) \to x_0,\ f(a(\lambda)) \to f(x_0) \quad \text{for } \lambda \to 0.$$

We suppose that f allows an expansion

(5) $$f(a(\lambda)) = f(x_0) + \lambda f_1 + \lambda^2 f_2 + \dots,\ f_i \in \mathbb{R}.$$

That means, the relations

(6) $$\frac{1}{\lambda}(f(a(\lambda)) - f(x_0)) \longrightarrow f_1,$$

$$\frac{1}{\lambda}\left(\frac{1}{\lambda}(f(a(\lambda)) - f(x_o)) - f_1\right) \to f_2, \ldots$$

hold for $\lambda \to 0$.

If x_o is a local minimal element of A with respect of f (in the usual sense) then obvious necessary conditions are given by

$$f_1 \geq 0; \quad f_2 \geq 0, \text{ if } f_1 = 0; \quad \ldots$$

Under the assumption that f is suffienctly smooth the variations $f_1, f_2, \ldots$ depend only on the directions $x_1, x_2, \ldots$, defined by

(7) $$\frac{a(\lambda) - x_o}{\lambda} \to x_1, \; \frac{1}{\lambda}\left(\frac{1}{\lambda}(a(\lambda) - x_o) - x_1\right) \to x_2, \ldots \quad .$$

Directions generated by certain classes of curves in A form so-called variational sets. The needed smoothness conditions on f are reflected by certain differentiability properties of f. This suggests the notion of variational derivatives.

We shall now suppose throughout this section that $(X, \mathcal{U})$ and $(Y, \hat{\mathcal{U}})$ are real topological vector spaces with filters of neighborhoods $\mathcal{U}(x)$ and $\hat{\mathcal{U}}(y)$ for $x \in X$, $y \in Y$. $\mathcal{V}$ and $\hat{\mathcal{V}}$ will denote additional topologies on X and Y, respectively, which are not necessarily compatible with the linear space structure on X and Y. Let $A \subset X$ be a subset, and let $x_o, \ldots, x_{m-1} \in X$ be elements of X, $m \in \mathbb{N}$.

Definition. The sets

$$K^m(A) := K^m_{\mathcal{U},\mathcal{V}}(A) := K_{\mathcal{U},\mathcal{V}}(A; x_o, \ldots, x_{m-1})$$
$$:= \{x \in X \mid \exists U \in \mathcal{U}(x_o) \;\; \exists V \in \mathcal{V}(x) \;\; \exists \eta > 0 \;\; \forall \lambda \in (0, \eta)$$
$$\left(\sum_{j=0}^{m-1} \lambda^j x_j + \lambda^m V\right) \cap U \subset A\},$$

$$K^m[A] := K^m_{\mathcal{U},\mathcal{V}}[A] := K_{\mathcal{U},\mathcal{V}}[A; x_o, \ldots, x_{m-1}]$$
$$:= \{x \in X \mid \forall U \in \mathcal{U}(x_o) \;\; \forall V \in \mathcal{V}(x) \;\; \forall \eta > 0 \;\; \exists \lambda \in (0, \eta)$$
$$\left(\sum_{j=0}^{m-1} \lambda^j x_j + \lambda^m V\right) \cap U \cap A \neq \phi\}$$

are called *variational sets of order* m *of* A *at* x_o (with respect to the points $x_1, \ldots, x_{m-1}$ and the topologies $\mathcal{U}, \mathcal{V}$).

The following properties follow easily from the definition.

P 1. $K^m(A)$ is $\mathcal{V}$-open, $K^m[A]$ is $\mathcal{V}$-closed.

P 2. $\forall U \in \mathcal{U}(x_o)$ $\quad K^m(A \cap U) = K^m(A)$ and $K^m[A \cap U] = K^m[A]$.

P 3. $\forall A_i \subset X$ $\quad K^m(A_1) \cap K^m[A_2] \subset K^m[A_1 \cap A_2]$.

P 4. If for every $\alpha > 0$ the mapping $x \to \alpha x$ is $\mathcal{V}$- continuous on X, then

$$x \in K(A;x_o,\dots,x_{m-1}),\ \alpha > 0 \Rightarrow \alpha^m x \in K(A;x_o,\ \alpha x_1,\dots,\alpha^{m-1}x_{m-1}),$$

$$x \in K[A;x_o,\dots,x_{m-1}],\ \alpha > 0 \Rightarrow \alpha^m x \in K[A;x_o,\ \alpha x_1,\dots,\alpha^{m-1}x_{m-1}].$$

Thus $K(A;\, x_o)$ and $K\,[A;x_o]$ are cones. In case $m > 1$ this conclusion fails in general.

P 5. Let $\mathcal{U}_1$ and $\mathcal{V}_1$ be topologies on X, which are finer than $\mathcal{U}$ and $\mathcal{V}$, respectively. Then

$$K^m_{\mathcal{U},\mathcal{V}}(A) \subset K^m_{\mathcal{U}_1,\mathcal{V}_1}(A) \subset K^m_{\mathcal{U}_1,\mathcal{V}_1}[A] \subset K^m_{\mathcal{U},\mathcal{V}}[A].$$

P 6. Let $(X_i,\mathcal{U}_i)$ be topological vector spaces with additional topologies $\mathcal{V}_i$, and let $A_i \subset X_i$ be subsets, i=1,2. If we denote by $\mathcal{U}_1 \times \mathcal{U}_2$ and $\mathcal{V}_1 \times \mathcal{V}_2$ the product topologies on $X_1 \times X_2$, then

$$K^m_{\mathcal{U}_1 \times \mathcal{U}_2,\mathcal{V}_1 \times \mathcal{V}_2}(A_1 \times A_2) = K^m_{\mathcal{U}_1,\mathcal{V}_1}(A_1) \times K^m_{\mathcal{U}_2,\mathcal{V}_2}(A_2),$$

$$K^m_{\mathcal{U}_1 \times \mathcal{U}_2,\mathcal{V}_1 \times \mathcal{V}_2}[A_1 \times A_2] \subset K^m_{\mathcal{U}_1,\mathcal{V}_1}[A_1] \times K^m_{\mathcal{U}_2,\mathcal{V}_2}[A_2].$$

In case m=1 it is possible to describe variational sets geometrically. For example, if A is starshaped at x_o, then

(8) $$K_{\mathcal{U},\mathcal{U}}(A;x_o) = \text{cone}\ (\text{int}_{\mathcal{U}}A - x_o).$$

If A is locally starshaped at x_o, then

(9) $$K_{\mathcal{U},\mathcal{V}}[A;x_o] = \text{cl}_{\mathcal{V}}\text{cone}\ (A - x_o).$$

It should be noted that (9) is valid if and only if $A-x_o \subset K_{\mathcal{U},\mathcal{V}}[A;x_o]$.

As we shall see later, the following notion of variational derivative will be useful in describing variational sets analytically.

Let $A \subset B \subset X$ be subsets, and let $x_o,\dots,x_{m-1} \in X$, $y_1,\dots,y_{m-1} \in Y$ be elements, $m \in \mathbb{N}$. We will first present the definition of differentiability of an arbitrary function $f : B \to Y$.

Definition. An element $y \in Y$ is called a

$K_{\mathcal{U},\mathcal{V}}[A;x_o,\dots,x_{m-1}]$-$(\hat{\mathcal{U}},\hat{\mathcal{V}})$-*differential of* f *at* x_o *in the direction* x *with respect to* $y_1,\dots y_{m-1}$, if

(a) $x \in K^m[A]$,

(b) $y_j \in f^{(j)}[A;x_o,\dots,x_{j-1};y_1,\dots,y_{j-1}](x_j)$ for $j=1,2,\dots,m-1$ (if $m \geq 2$),

(c) $\forall \hat{U} \in \hat{\mathcal{U}}(y_o)$ $\forall \hat{V} \in \hat{\mathcal{V}}(y)$ $\forall \hat{\eta} > 0$

$\exists U \in \mathcal{U}(x_o)$ $\exists V \in \mathcal{V}(x)$ $\exists \eta > 0$ $\forall \lambda \in (0, \min(\eta, \hat{\eta}))$

$$f\left(\left(\sum_{j=0}^{m-1} \lambda^j x_j + \lambda^m V\right) \cap U \cap A\right) \subset \left(\sum_{j=0}^{m-1} \lambda^j y_j + \lambda^m \hat{V}\right) \cap \hat{U},$$

where $y_o := f(x_o)$.

The set of all $K^m[A]$-differentials is denoted by

$$f^{(m)}[A](x) := f^{(m)}_{\mathcal{U},\mathcal{V},\hat{\mathcal{U}},\hat{\mathcal{V}}}[A;x_o,\dots,x_{m-1};y_1,\dots,y_{m-1}](x).$$

<u>Definition.</u> A mapping $f_A^{(m)} : K^m[A] \to Y$ is called a $K^m[A]$-*derivative of f at x_o with respect* to $y_1,\dots,y_{m-1}$, if $f_A^{(m)}(x) \in f^{(m)}[A](x)$ for each $x \in K^m[A]$.

<u>Remarks.</u>

1. In case $y_1 = \dots = y_{m-1} = \theta$ we refer to $f_A^{(m)}$ as a $K^m[A]$-derivative of f at x_o with respect to θ.

2. If the topology $\hat{\mathcal{V}}$ is separated, then $f^{(m)}[A](x)$ is a singleton. In this case $K^m[A]$-differentiability of f at x_o with respect to θ means, in particular, that $f^{(j)}[A](x_j)$ exists and is equal to θ for $j=1,2,\dots,m-1$.

3. We get the notion of $K^m(A)$-differentiability, if we replace $K^m[A]$ by $K^m(A)$.

We have the following (easily verified) properties for $K^m[A]$-derivatives.

<u>P 7.</u> If for every $\alpha > 0$ the mappings $x \to \alpha x$ and $y \to \alpha y$ are $\mathcal{V}$- and $\hat{\mathcal{V}}$-continuous on X and Y, respectively, then

$$\alpha^m f^{(m)}[A;x_o,\dots,x_{m-1};y_1,\dots,y_{m-1}](x) \subset f^{(m)}[A;x_o,\alpha x_1,\dots,\alpha^{m-1}y_{m-1};\alpha y_1,\dots,\alpha^{m-1}y_{m-1}](\alpha^m x)$$

for $x \in K^m[A]$, $\alpha > 0$.

In particular, when $\hat{\mathcal{V}}$ is separated, then $f'[A;x_o] : K[A;x_o] \to Y$ is positively homogenous.

<u>P 8.</u> Let $\mathcal{U}_1$ and $\mathcal{V}_1$ be topologies on X which are finer than $\mathcal{U}$ and $\mathcal{V}$, respectively, and let $\hat{\mathcal{U}}_1$ and $\hat{\mathcal{V}}_1$ be topologies on Y which are coarser than $\hat{\mathcal{U}}$ and $\hat{\mathcal{V}}$, respectively. Then

$$f^{(m)}_{\mathcal{U},\mathcal{V},\hat{\mathcal{U}},\hat{\mathcal{V}}}[A](x_m) \subset f^{(m)}_{\mathcal{U}_1,\mathcal{V}_1,\hat{\mathcal{U}}_1,\hat{\mathcal{V}}_1}[A](x_m)$$

provided that $x_j \in K^j_{\mathcal{U},\mathcal{V}}[A] \cap K^j_{\mathcal{U}_1,\mathcal{V}_1}[A]$ for $j=1,2,\dots,m$.

<u>P 9.</u> The chain rule holds without any additional assumption on the topologies. A more precise statement has been given in [9].

P 10. Suppose that Y is the direct topological product of Y_1 and Y_2, i. e. $Y = Y_1 \times Y_2$, $\hat{U} = \hat{U}_1 \times \hat{U}_2$, $\hat{V} = \hat{V}_1 \times \hat{V}_2$. Let (f_1, f_2) denote the canonical decomposition of f with respect to $Y_1 \times Y_2$. Suppose that $f^{(m)}_{i,A}$ are $K^m[A]$ - $(\hat{U}_i, \hat{V}_i)$-derivatives of f_i at x_o with respect to $y_{1,i}, \ldots, y_{m-1,i}$ for $i=1,2$. Then $f^{(m)}_A : K^m[A] \to Y$, defined by $f^{(m)}_A := (f^{(m)}_{1,A}, f^{(m)}_{2,A})$, is a $K^m[A]$ - $(\hat{U}, \hat{V})$ derivative of f at x_o with respect to $(y_{1,1}, y_{1,2}), \ldots, (y_{m-1,1}, y_{m-1,2})$.

P 11. We consider the special case that X and Y are normed linear spaces (with normtopologies U and $\hat{U}$, respectively), and we denote $K^m_{U,U}[A]$ by $T^m_U[A]$. Let f be m-times Frêchet-differentiable at x_o. Then f is $T^m_U[A]$-differentiable at x_o in any direction $x \in T^m_U[A]$ and

$$\frac{1}{m!} \left. \frac{d^m f\left(\sum_{j=0}^{m-1} \lambda^j x_j + \lambda^m x\right)}{d\lambda^m} \right|_{\lambda=0}$$

exists and is equal to $f^{(m)}_{U,U,\hat{U},\hat{U}}[A](x)$.

By the chain rule we obtain

$$f'[A;x_o](x_1) = f^o_x(x_1),$$

$$f''[A;x_o,x_1](x_2) = f^o_x(x_2) + \frac{1}{2} f^o_{xx}(x_1,x_1),$$

$$f'''[A;x_o,x_1,x_2](x_3) = f^o_x(x_3) + f^o_{xx}(x_1,x_2) + \frac{1}{6} f^o_{xxx}(x_1,x_1,x_1),$$

where $f^o_x, f^o_{xx}, \ldots$ denote the Frêchet-derivatives at x_o.

The following *invariance property* shows that the notion of variational derivatives is well adapted to the notion of variational sets.

Theorem 1. Let $f^{(m)}_A : K^m[A] \to Y$ be a $K^m[A]$-derivative of f at x_o with respect to $y_{1,\ldots,m-1}$. Then

(10) $\quad K_{U,V}[A;x_o,\ldots,x_{m-1}] = (f^{(m)}_A)^{-1}(K_{\hat{U},\hat{V}}[f(A);y_o,\ldots,y_{m-1}])$,

where $y_o := f(x_o)$.

Proof. The inclusion $\supset$ is trivial. To reverse the inclusion we consider $x \in K^m[A]$, $y := f^{(m)}_A(x) \in f^{(m)}[A](x)$, $\hat{V} \in \hat{U}(y_o)$ $\hat{\hat{V}} \in \hat{V}(y)$ and $\hat{\eta} > 0$. From the definition it follows that there are $U \in U(x_o)$, $V \in V(x)$ and $\eta > 0$, such that $f((\sum_{j=0}^{m-1} \lambda^j x_j + \lambda^m V) \cap U \cap A) \subset (\sum_{j=0}^{m-1} \lambda^j y_j + \lambda^m \hat{V}) \cap \hat{U}$ for all $\lambda \in (0,\eta_o)$, $\eta_o := \min(\eta,\hat{\eta})$. $x \in K^m[A]$ means that there exist $\lambda_o \in (0,\eta_o)$

and $v_o \in V$ such that $a_o := \sum_{j=0}^{m-1} \lambda_o^j x_j + \lambda_o^m v_o \in U \cap A$ and

$f(a_o) \in (\sum_{j=0}^{m-1} \lambda^j y_j + \lambda^m \hat{V}) \cap \hat{U} \cap f(A)$. □

Remark. In particular, if $\hat{\mathcal{V}}$ is separated, then (10) takes the following form.

$$K_{\mathcal{U},\mathcal{V}}[A;x_o,\dots,x_{m-1}] = (f_A^{(m)})^{-1} K_{\hat{\mathcal{U}},\hat{\mathcal{V}}}[f(A);f_o,\dots,f_{m-1}], \tag{11}$$

where f_j denotes the $K^j[A]$-differential $f^{(j)}[A](x_j)$ for $j=1,2,\dots,m-1$ and $f_o = f(x_o)$.

As an immediate consequence of theorem 1 we note

Collary 2. Let $Q \subset Y$ be an arbitrary subset. Let $f_B^{(m)}: K^m[B] \to Y$ be a $K^m[B]$-derivative of f at x_o with respect to $y_1,\dots,y_{m-1}$. Then

$$K^m[A] \subset (f_B^{(m)})^{-1}(K[f(A);y_o,\dots,y_{m-1}]) \tag{12}$$

and

$$K^m[f^{-1}(Q)] \subset (f_B^{(m)})^{-1}(K[Q;y_o,\dots,y_{m-1}]). \tag{13}$$

The reverse inclusion holds for $K^m(B)$-derivatives.

Theorem 3. Let $Q \subset Y$ be an arbitrary subset. Let $f_B^{(m)} : K^m(B) \to Y$ be a $K^m(B)$-derivative of f at x_o with respect to $y_1,\dots,y_{m-1}$. Then

$$(f_B^{(m)})^{-1}(K(Q\,;\,y_o,\dots,y_{m-1})) \subset K^m(f^{-1}(Q)). \tag{14}$$

An analytic description of subsets of $K^m[f^{-1}(Q)]$ is possible if we impose additional restrictions on X,Y and f.

Definition. f is said to satisfy conditions (G1) and (G2) at x_o, respectively, if

(G1) (a) X,Y are Banach spaces (with norm topologies $\mathcal{U} = \mathcal{V}$, $\hat{\mathcal{U}} = \hat{\mathcal{V}} =: \mathcal{W}$)

(b) $x_o \in \text{int } B$,

(c) f is continuous in a neighborhood of x_o and f is continuously Gateaux-differentiable at x_o,

(d) f is m-times Fréchet-differentiable at x_o.

(G2) the Fréchet-derivative $f_x(x_o)$ is surjective.

The following theorem extends the classical theorem of Lyusternik [10].

Theorem 4. Let $Q \subset Y$ be an arbitrary subset. Suppose that f satisfies conditions (G1) and (G2) at x_o. If $f^{(j)}[B;x_o,\dots,x_{j-1}](x_j) = \theta$ for $j=1,2,\dots,m-1$, then

$$(f_B^{(m)})^{-1}(T_{\mathcal{W}}[Q;f(x_o)]) = T_{\mathcal{V}}[f^{-1}(Q);x_o,\dots x_{m-1}]. \tag{15}$$

Now we are able to obtain the following necessary conditions of order m for the Problem (1).

Theorem 5. Let $x_o \in A$ be a Q-minimal element of A with respect to f. Let $f_B^{(m)} : K[B;x_o,\dots,x_{m-1}] \to Y$ be a $K^m[B]$-derivative of f at x_o with respect to θ.
Then

(16) $K(Q,\theta) \cap K[f(A) - f(x_o);\theta] = \phi$

and

(17) $K(Q,\theta) \cap f_B^{(m)}(K[A;x_o,\dots x_{m-1}]) = \phi$

Proof. $K[Q \cap (f(A) - f(x_o));\theta] = \phi$ follows immediately from (1), and (P3) shows that (16) holds. By Collary 2 we see that $f_B^{(m)}(K^m[A]) \subset$

$K[f(A);f(x_o),\theta,\dots,\theta] = K[f(A);f(x_o)] = K[f(A) - f(x_o);\theta]$. □

Remark. It should be noted that (16) is a first order condition in Y.

In case $K(Q,\theta) = \phi$ the necessary conditions (16) and (17) are clearly trivial, and it is thus of interest to refine the results of Theorem 5. For this purpose we assume just as in P 10 that Y is the direct topological product of Y_2 and Y_2; i. e. $Y = Y_1 \times Y_2$, $\hat{U} = \hat{U}_1 \times \hat{U}_2, \hat{V} = \hat{V}_1 \times \hat{V}_2$. Let (f_1,f_2) denote the canonical decomposition of f with respect to $Y_1 \times Y_2$. Let $Q_i \subset Y_i$ be subsets.

Theorem 6. Let $x_o \in A$ be a $Q_1 \times Q_2$-minimal element of A with respect to (f_1,f_2).

Suppose that f_2 satisfies conditions (G1), (G2) at x_o, and assume that $f_2^{(j)}[B;x_o,\dots x_{j-1}](x_j) = \theta$ for $j=1,2,\dots,m-1$.
Let $f_{1,B}^{(m)} : K_{U,\tilde{V}}[B;x_o,\dots,x_{m-1}] \to Y_1$ be a $K^m[B]$- $(\hat{U}_1,\hat{V}_1)$-derivative of f_1 at x_o with respect to θ.

Then if U is finer than $\tilde{V}$

(18) $(K_{\hat{U}_1,\hat{V}_1}(Q_1;\theta) \times T_{\hat{U}_2}[Q_2;\theta]) \cap (f_{1,B}^{(m)},f_{2,B}^{(m)})(K^m_{U,\tilde{V}}(A)) = \phi.$

Proof. We define $B_i := f_i^{-1}(f_i(x_o) + Q_i)$, $i=1,2$. Since x_o is a $(Q_1 \times Q_2)$-minimal element of A with respect to (f_1,f_2) we have $A \cap B_1 \cap B_2 = \phi$. Hence, it follows that $K^m_{U,\tilde{V}}[A \cap B_1 \cap B_2] = \phi$ and by P3 $K^m_{U,\tilde{V}}(A) \cap K^m_{U,\tilde{V}}(B_1) \cap K^m_{U,\tilde{V}}[B_2] = \phi$.
Theorem 3 shows that $(f_{1,B}^{(m)})^{-1}(K^m(Q_1;\theta)) \subset K^m(B_1)$. By theorem 4 it follows that $(f_{2,B}^{(m)})^{-1}(T[Q_2;\theta]) = T^m_U[B_2]$. Finally, $T^m_U[B_2] \subset K_{U,\tilde{V}}[B_2]$, completing the proof of (18). □

Remarks.

1. It is worth noting that (17) and (18) are by virtue of P2 *local* necessary conditions (with respect to $\mathcal{U}$).

2. As we noted in section 1 necessary conditions of the form (16), (17) or (18) imply maximum principles (of order m) if certain hypotheses on the data of the problems are satisfied.

The results presented here form a framework within which it is possible to obtain higher-order conditions for various types of optimization problems. Since in this paper we are mainly interested in applications to semi-infinite programming problems, we have confined ourselves to the essentials of a unified higher-order theory. A more detailed discussion of the general case can be found in [8],[9].

3. Necessary Conditions for Semi-Infinite Programming Problems

In this section we specialize the preceeding results to semi-infinite programming problems. For this purpose we consider the maximum-function which plays a central role in many applications, especially, as we have seen in example 4, in semi-infinite programming problems. We investigate various conditions, which guarantee, that the maximum function is $K^m[A]$-differentiable.

Now let $(X,\mathcal{U})$ be a topological vector space with an additional topology $\mathcal{V}$, let Y be the real line $\mathbb{R}$ (with the usual topology $\hat{\mathcal{U}} = \hat{\mathcal{V}}$) and let $(S,\mathcal{W})$ be a compact Hausdorff space. Let $A \subset B \subset X$ be subsets and $F : B \times S \to \mathbb{R}$ a mapping. Let us assume that $f : B \to \mathbb{R}$, defined by $f(x) = \max_{s\in S} F(x,s)$ exists. Let $x_o \in B$, $x_1,\dots,x_m \in X$ be elements, $m \in \mathbb{N}$.

Theorem 7. Suppose that the following hypotheses (D1)-(D4) are satisfied.

(D1) For all $s\in S$ the $K^m[A]$-differential
$f_m(s) := f^{(m)}[A;x_o,\dots,x_{m-1};s](x_m)$
(and $f_j(s) := f^{(j)}[A;x_o,\dots,x_{j-1};s](x_j)$ for $j=1,2,\dots,m-1$) of $F(\cdot,s) : B \to \mathbb{R}$ at x_o in direction x_m exists. The mappings $f_j : S \to \mathbb{R}$ are upper semicontinuous for $j=0,1,\dots,m$, where f_o is defined by $f_o(s) := F(x_o,s)$.

(D2) $F : B \times S \to \mathbb{R}$ is S-equi-$K^m[A]$-differentiable at x_o in direction x_m, i. e.

$$\forall \varepsilon > 0 \quad \exists U \in \mathcal{U}(x_o) \quad \exists V \in \mathcal{V}(x_m) \quad \exists \eta > 0$$
$$\forall u \in U \quad \forall v \in V \quad \forall (\lambda \in (0,\eta) : u = \sum_{j=0}^{m-1} \lambda^j x_j + \lambda^m v \in A)$$

$\frac{1}{\lambda}(\ldots\frac{1}{\lambda}(F(u,s) - f_o(s)) - \ldots - f_{m-1}(s)) - f_m(s) < \varepsilon$

for all $s \in S$.

(D3) For $j=1,2,\ldots,m-1$ (if $m \geq 2$) the $K^{(j)}[A]$-differentials $f^{(j)}[A;x_o,\ldots,x_{j-1}](x_j)$ exists and are equal to $f_j := \max_{s\in E_{j-1}} f_j(s)$, where

$E_o := E(x_o)$, $E_j := \{s \in E_{j-1} | f_j = f_j(s)\}$ and

$E(x) := \{s \in S | f(s) = F(x,s)\}$.

(D4) $\exists c \geq 0 \quad \forall \varepsilon > 0 \quad \exists U \in \mathcal{U}(x_o) \quad \exists V \in \mathcal{V}(x_m) \quad \exists \eta > 0$

$\forall u \in U \quad \forall v \in V \quad \forall (\lambda \in (0,\eta) : u = \sum_{j=0}^{m-1} \lambda^j y_j + \lambda^m v \in A)$

$\exists s \in E(u)$

(19) $\frac{1}{\lambda}(\ldots\frac{1}{\lambda}(f_o(s) - f_o) + \ldots + (f_{m-1}(s) - f_{m-1})) \leq c\varepsilon$

Then f is $K^m[A]$-differentiable at x_o in the direction x_m, and

$$f_m := f^{(m)}[A;x_o,\ldots,x_{m-1}](x_m) = \max_{s \in E_{m-1}} f_m(s).$$

The proof proceeds along the same lines as in case $\mathcal{U} = \mathcal{V}$ (see [8], pp. 560).

<u>Remarks.</u>

1. In case $m=1$ condition (D3) is empty, and condition (D4) is satisfied with $c=0$ since $f_o(s) \leq f_o$ for all $s \in S$.

2. In case $m \geq 2$ condition (D4) with $c=0$ is implied by

(D4') $\exists U \in \mathcal{U}(x_o) \quad \exists V \in \mathcal{V}(x_m) \quad \exists \eta > 0$

$\forall u \in U \quad \forall v \in V \quad \forall (\lambda \in (0,\eta) : u = \sum_{j=0}^{m-1} \lambda^j x_j + \lambda^m v \in A)$

$E(u) \subset E_{m-2}$.

It is clear that condition (D4') is satisfied for every $m \in \mathbb{N}$ if S is a finite set. This situation has been considered in [1] with respect to directional differentiability of f.

As an important special case of semi-infinite programming we now consider the problem of *best uniform approximation*. Let $X := C(S,\mathbb{R})$ denote the space of all real-valued continuous functions on the compact Hausdorff space S normed by the uniform norm $\|x\| = \max_{s\in S} |x(s)|$. Let

$$P_M(y) := \{x \in X \mid \exists U \in \mathcal{U}(x) \ \|x-y\| = \inf_{x\in M\cap U} \|x-y\| \}$$

denote the set of all elements of local best approximation of $y \in X$ by elements of the set $M \subset X$ (with respect to the norm $\|\cdot\|$), where $\mathcal{U}$ denotes the normtology on X.

<u>Theorem 8.</u> Let $M \subset X$ be a nonempty subset and let $y \in X$. Suppose that $x_o \in P_M(y)$ satisfies

(D4") $\exists U \in \mathcal{U}(x_o) \quad \forall u \in U \qquad E(u) \subset E(x_o)$,

where $E(x) := \{s \in S \mid \|x-y\| = |x_o(s) - y(s)|\}$.

Then if $r_o(s) := x_o(s) - y(s)$ and $\|r_o\| > 0$ the following necessary conditions hold.

(20) $$\max_{s \in E(x_o)} x_1(s) \operatorname{sgn} r_o(s) \geq 0 \qquad \text{for all } x_1 \in T[M;x_o]$$

and

(21) $$\max_{s \in E(x_o,x_1)} \operatorname{sgn} x_2(s) \operatorname{sgn} r_o(s) \geq 0 \quad \text{for all } x_2 \in T[M;x_o,x_1]$$

and for all $x_1 \in T[M;x_o]$ with $\max_{s \in E(x_o)} x_1(s) r_o(s) = 0$, where

$$E(x_o,x_1) := \{s \in E(x_o) \mid x_1(s) \operatorname{sgn} r_o(s) = 0\}.$$

<u>Proof.</u> If we define $F : X \times S \to \mathbb{R}$ by $F(x,s) = |x(s) - y(s)|$, then $f : X \to \mathbb{R}$ defined by $f(x) = \|x-y\| = \max_{s \in S} F(x,s)$ exists. Let $\mathcal{U} = \mathcal{V}$ and $\hat{\mathcal{U}} = \hat{\mathcal{V}}$ denote the normtopology on X and the usual topology on $\mathbb{R}$, respectively. It is easy verified that for any $x_1 \in X$ and for any $s \in S$

$$f_1(s) := \begin{cases} x_1(s) \operatorname{sgn} r_o(s) & , r_o(s) \neq 0 \\ |x_1(s)| & , r_o(s) = 0 \end{cases}$$

is the $T_{\mathcal{U}}[X;x_o]$-differential of $F(\cdot,s) : X \to \mathbb{R}$ at x_o in the direction x_1. Moreover, $F : X \times S \to \mathbb{R}$ is S-equi-$T_{\mathcal{U}}[X;x_o]$-differentiable at x_o in the direction x_1.

As a first consequence of Theorem 7, it follows that for any $x_1 \in X$ $f'(x_o)(x_1) = f_1 = \max_{s \in E(x_o)} x_1(s) r_o(s)$ is the $T_{\mathcal{U}}[X;x_o]$-differential of f at x_o in the direction x_1.

Applying Theorem 5 we obtain (20), since $K(Q,\theta) = Q$ for $Q := \{y \in \mathbb{R} \mid y < 0\}$.

(21) follows in the same way. As in the proof of (20) we first see that for any $x_1,x_2 \in X$ and for any $s \in S$

$$f_2(s) := \begin{cases} x_2(s) \operatorname{sgn} r_o(s) & , r_o(s) \neq 0 \\ x_2(s) \operatorname{sgn} x_1(s) & , r_o(s) = 0,\ x_1(s) \neq 0 \\ |x_2(s)| & , r_o(s) = x_1(s) = 0 \end{cases}$$

is the $T_{\mathcal{U}}[X;x_o,x_1]$-differential of $F(\cdot,s) : X \times S \to \mathbb{R}$ at x_o in the direction x_2.

Further it is easily verified that $F : X \times S \to \mathbb{R}$ is S-equi-$T_{\mathcal{U}}[X;x_o,x_1]$-differentiable at x_o in the direction x_2.

Finally, hypothesis (D4') is satisfied by hypothesis (D4"). Hence, by Theorem 7 and by Theorem 5 we conclude (21). □

Remarks.

1. For simplicity we have presented only necessary conditions of order m=1,2. But the proof shows that it is possible to derive conditions of order m > 2 without additional difficulties.

2. (20) is the well-known local Kolmogorov-condition.

The following simple example shows that the condition (21) is able to distinguish local best approximations from other points which satisfy the local Kolmogorov condition (20).

E 5. Let S be the interval [-1,+1], let y be defined by $y(s) = 1-8s+s^2$ and let $M = g(\mathbb{R})$, where g is defined by $g(a) := 2a^2g_o - 4ag_1$, $g_i(s) := s^i$. By Collary 2 we have

$$g'[\mathbb{R};a_o]\,(T[\mathbb{R};a_o]) \subset T[M;x_o]$$

and

$$g''[\mathbb{R};a_o,a_1]\,(T[\mathbb{R};a_o,a_1]) \subset T[M;x_o,x_1],$$

where

$$x_o = g(a_o) \text{ and } x_1 = g'[\mathbb{R};a_o](a_1).$$

Hence, for every local best approximation x_o, which satisfies the condition (D4") of Theorem 8 we have the necessary conditions

$$(22) \qquad \max_{s\in E(x_o)} (4a_o - 4s)a_1 \operatorname{sgn} r_o(s) \geq 0 \qquad \text{for all } a_1 \in \mathbb{R}$$

and

$$(23) \qquad \max_{s\in E(x_o,x_1)} [(4a_o - 4s)a_2 + 4a_1^2] \operatorname{sgn} r_o(s) \geq 0 \quad \text{for all } a_2 \in \mathbb{R}$$

and for all $a_1 \in \mathbb{R}$ with $\max_{s\in E(x_o)} (4a_o - 4s)a_1 \operatorname{sgn} r_o(s) = 0$.

Condition (22) is satisfied by $a_o = \pm 1$. In both cases $x_o = g(a_o)$ satisfies (D4"). But (23) holds only for $a_o = +1$. It is easy to see that indeed $x_o = g(+1)$ is the unique best approximation of y by elements of M.

As we have seen in the proof of Theorem 8 the supremum norm is $T[X;x_o]$-differentiable at any point $x_o \in X$ in any direction $x_1 \in X$. An analogous result does not hold for $T[X;x_o,x_1]$-differentiability. The assumption of (D4") (or D4') here is vital, as the following example shows.

E 6. Let S be the interval $[0,1]$ and $y=\theta$. Let $x_0, x_1, x_2 \in X$ be defined by $x_0(s) := 1-s^2$, $x_1(s) := \frac{2}{\alpha} s^{\alpha}, \alpha > 0$, and $x_2(s) := 1$. Then $f'(x_0)(x_1) = f_1 = \max_{s \in E(x_0)} x_1(s) \operatorname{sgn} x_0(s) = x_1(0) = 0$ and $f_2 := \max_{s \in E(x_0, x_1)} x_2(s) \operatorname{sgn} x_0(s) = x_2(0) = 1$ exist. But it is immediately verified that

$$f''(x_0,x_1)(x_1) = \begin{cases} f_2, & \alpha > 1 \\ f_2 + 1, & \alpha = 1 \end{cases}$$

and that $f''(x_0,x_1)(x_2)$ does not exist if $0 < \alpha < 1$.

As a detailed investigation of this and other examples shows, it is useful in discussing the differentiability of the maximum function to consider differentials of F with respect to both variables x and s.

Let us now turn back to the general situation of this section.

Suppose that $S \subset R \subset Z$ and $F : B \times R \to \mathbb{R}$, where $(Z, \mathcal{W})$ is a real topological vector space.

Theorem 9. Suppose that the following hypotheses (H1)-(H4) hold.

(H1) $x_0 \in \operatorname{int}_{\mathcal{U}} B$,

(H2) For $j=1,2,\ldots,m$:

(a) $S_{j-1} \neq \phi$

(b) For every $(s_0,\ldots,s_{j-1}) \in S_{j-1}$ and for every $s_j \in \overline{T}_{\mathcal{W}}(S;s_0,\ldots,s_{j-1})$ the $K_{\mathcal{U}\times\mathcal{W},\mathcal{V}\times\mathcal{W}}[B\times S;(x_0,s_0),\ldots,(x_{j-1},s_{j-1})]$-differential $F_j(s_0,\ldots,s_j) := F^{(j)}[B\times S;(x_0,s_0),\ldots,(x_{j-1},s_{j-1})](x_j,s_j)$ of $F : B \times R \to \mathbb{R}$ at (x_0,s_0) in the direction (x_j,s_j) exists.

(c) $F_j := \sup_{(s_0,\ldots,s_{j-1}) \in S_{j-1}} \ \sup_{s_j \in \overline{T}_{\mathcal{W}}(S;s_0,\ldots,s_{j-1})} F_j(s_0,\ldots,s_j)$ exists.

(d) $S_j := \{(s_0,\ldots,s_j) \in S_{j-1} \times \overline{T}^j_{\mathcal{W}}(S) \mid F_j = F_j(s_0,\ldots,s_j)\}$.

(H3) For $j=1,2,\ldots,m-1$ (if $m \geq 2$) the $K^{(j)}_{\mathcal{U},\mathcal{V}}[B]$-differentials $f^{(j)}[B;x_0,\ldots,x_{j-1}](x_j)$ exist and are equal to f_j.

(H4) For every $(s_0,\ldots,s_{m-1}) \in S_{m-1}$:

(a) $F_m(s_0,\ldots,s_{m-1},\cdot) : \overline{T}_{\mathcal{W}}(S;s_0,\ldots,s_{m-1}) \to \mathbb{R}$ is lower semicontinuous.

(b) $\forall \varepsilon > 0 \quad \exists U \in \mathcal{U}(x_o) \quad \exists V \in \mathcal{V}(x_m) \quad \exists \eta > 0$

$$\forall u \in U \quad \forall v \in V \quad \forall (\lambda \in (0,\eta) : u = \sum_{j=0}^{m-1} \lambda^j x_j + \lambda^m v \in B)$$

$$\exists t \in \overline{T}_{\mathcal{W}}(S; s_o, \ldots, s_{m-1}) \qquad \exists s = \sum_{j=0}^{m-1} \lambda^j s_j + \lambda^m t \in E(u)$$

$$\Delta^m F(u,s) - F_m(s_o \ldots, s_{m-1}, t) < \varepsilon ,$$

where

$$\Delta^m F(u,s) := \frac{1}{\lambda}(\ldots\frac{1}{\lambda}(F(u,s) - F(x_o,s_o))\ldots - F_{m-1}(s_o,\ldots,s_{m-1}))$$

denotes the difference quotient of order m.

Then f is $K^m_{\mathcal{U},\mathcal{V}}[B]$-differentiable at x_o in the direction x_m and $f^{(m)}[B; x_o, \ldots, x_{m-1}]\,(x_m) = F_m$.

Proof. First we note that for every $(s_o,\ldots,s_{m-1}) \in S_{m-1}$

$$\Delta^m F(x,s) = \frac{1}{\lambda}(\ldots\frac{1}{\lambda}(F(x,s) - F(x_o,s_o)) - \ldots - F_{m-1})$$

$$\leq \frac{1}{\lambda}(\ldots\frac{1}{\lambda}(f(x) - f_o) - \ldots - f_{m-1}) =: \Delta^m f(x)$$

for all $(x,s) \in B \times S$, where now f_i denote the differentials of f. Equality holds for $(x,s) \in B \times E(x)$. Because of (H1) we have $K_{\mathcal{U}\times\mathcal{W},\mathcal{V}\times\mathcal{W}}[B \times S] = K_{\mathcal{U},\mathcal{V}}[B] \times T[s]$.

Let $\varepsilon > 0$ be given.

1. Let $(s_o,\ldots,s_{m-1}) \in S_{m-1}$ and $s_m \in \overline{T}^m(S) := cl_{\mathcal{W}} T_{\mathcal{W}}(S;, s_o, \ldots, s_{m-1})$ be arbitrary elements. Since $F_m(s_o,\ldots,s_{m-1},\cdot)$ is lower semicontinuous there is an element $s \in T^m(S)$ such that $F_m(s_o,\ldots,s_m) - F_m(s_o,\ldots,s_{m-1},s) < \frac{\varepsilon}{2}$. By (H2) we have:

$\exists U \in \mathcal{U}(x_o) \quad \exists V \times W \in \mathcal{V}(x_m) \times \mathcal{W}(s) \quad \exists \eta > 0$

$$\forall u \in U \quad \forall z \in Z \quad \forall (v,w) \in V \times W \quad \forall (\lambda \in (0,\eta) : (u,z) = \sum_{j=0}^{m-1} \lambda^j (x_j,s_j) + \lambda^m (v,w) \in B \times S)$$

$$F_m(s_o,\ldots,s_{m-1},s) - \Delta^m F(u,z) < \varepsilon/2.$$

Taking into account that $s \in T^m(s)$ we obtain

$\exists U \in \mathcal{U}(x_o) \quad \exists V \in \mathcal{V}(x_m) \quad \exists \eta > 0$

$$\forall u \in U \quad \forall v \in V \quad \forall (\lambda \in (0,\eta) : u = \sum_{j=0}^{m-1} \lambda^j x_j + \lambda^m v \in B)$$

$F_m(s_o,\ldots,s_{m-1},s) < \Delta^m f(u) + \varepsilon/2$, hence

$F_m(s_o,\ldots,s_{m-1},s_m) < \Delta^m f(u) + \varepsilon$.

Making use of this fact we prove indirectly that

$\exists U_1 \in \mathcal{U}(x_o) \quad \exists V_1 \in \mathcal{V}(x_m) \quad \exists \eta_1 > 0$

$$\forall u \in U_1 \quad \forall v \in V_1 \quad \forall (\lambda \in (0,\eta_1) : u = \sum_{j=0}^{m-1} \lambda^j x_j + \lambda^m v \in B)$$

$F_m < \Delta^m f(u) + \varepsilon$.

2. Hypothesis (H4) shows that there are $U_2 \in \mathcal{U}(x_o)$, $V_2 \in \mathcal{V}(x_2)$ and $\eta_2 > 0$ such that

$$\forall u \in U_2 \quad \forall v \in V_2 \quad \forall (\lambda \in (0,\eta_2) : u = \sum_{j=0}^{m-1} \lambda^j x_j + \lambda^m v \in B)$$

$\Delta^m f(u) < F_m + \varepsilon$. □

Remarks.

1. By similar arguments it is easy to prove that (H4)(b) may be replaced by the following assumption on the differentiability of the mapping $E : B \to 2^S$.

$$\text{(H4)(b')} \quad \exists (s_o,\ldots,s_{m-1}) \in S_{m-1} \quad \exists s_m \in \overline{T}(S;s_o,\ldots,s_{m-1})$$
$$\forall W \in \mathcal{W}(s_m) \quad \exists U \in \mathcal{U}(x_o) \quad \exists V \in \mathcal{V}(x_m) \quad \exists \eta > 0$$
$$\forall u \in U \quad \forall v \in V \quad \forall (\lambda \in (0,\eta) : u = \sum_{j=0}^{m-1} \lambda^j x_j + \lambda^m v \in B) \ \exists s \in E(u)$$
$$\frac{1}{\lambda}(\ldots\frac{1}{\lambda}(s-s_o) - \ldots - s_{m-1}) \in W.$$

2. In case m=1,2 similar results concerning the directional differentiability of the maximum function are obtained in [2],[3],[4] and [5].

3. The main difficulty in proving the existence of the $K^m[B]$-differentials of the maximum function consists in verifying that hypothesis (H4)(b) holds. In special cases this hypothesis is implied by assumptions which naturally arise in connection with the calculation of $F_o, F_1, F_2, \ldots$

4. It is not difficult to derive analogous results under the assumption that F is $K_{\mathcal{U}\times\mathcal{W},\mathcal{V}\times\mathcal{W}'}[B \times S]$-differentiable, where $\mathcal{W}'$ denotes a second topology on Z. In this case we have to replace $\overline{T}_{\mathcal{W}}(s)$ by $\overline{K}_{\mathcal{W},\mathcal{W}'}(S)$.

As a first application we derive necessary conditions for the minimum norm problem in Hilbert space. We suppose that $X = Z = A = B = R := H$ is a Hilbert space with real inner product $\langle\cdot,\cdot\rangle$, and we choose $\mathcal{U} = \mathcal{V} = \mathcal{W}$ as the induced normtopology.

From the Hahn-Banach-theorem we have $f(x) := \|x\| = \max_{s\in S} F(x,s)$, where $F(x,s) := \langle x,s\rangle$ and $S := \{s \in H \mid \langle s,s\rangle - 1 \leq 0\}$. It is immediately verified that for every $x_o,x_1,x_2 \in H$ and for every $s_o,s_1,s_2 \in H$ the $T[B \times S]$-differentials of F exists and are equal to

$$F_1(s_o,s_1) = \langle x_o,s_1\rangle + \langle x_1,s_o\rangle ,$$

$$F_2(s_o,s_1,s_2) = \langle x_o,s_2\rangle + \langle x_1,s_1\rangle + \langle x_2,s_o\rangle .$$

Moreover, when $\|s_o\| = 1$, then we have

$$\overline{T}(S;s_o) = T[S;s_o] = \{s \in H \mid \langle s_o,s\rangle \leq 0\} \text{ and}$$

$$\overline{T}(S;s_o,s_1) = T[S;s_o,s_1] = \{s \in H \mid 2 <s_o,s> + <s_1,s_1> \leq 0\}$$

for all $s_1 \in H$, which satisfy $<s_o,s_1> = 0$.

Let us now assume that $x_o \neq \theta$. It follows at once that $S_o := E(x_o) = \{s_o\}$, $s_o := x_o/\|x_o\|$. Thus we have

$$F_1 = \sup_{s_o \in S_o} \sup_{s_1 \in T[S;s_o]} (<x_o,s_1> + <x_1,s_o>) = <x_1,s_o>,$$

$$S_1 = \{(s_o,s_1) \in S_o \times T[S;s_o] \mid <x_o,s_1> = 0\} \neq \phi$$

and $\max_{s_2 \in T[S;s_o,s_1]} <x_o,s_2> = -\frac{1}{2} <s_1,s_1> \|x_o\|$ exists for every $(s_o,s_1) \in S_1$. The quadratic maximum problem

$$\max_{\substack{s_1 \in T[S;s_o] \\ <x_o,s_1>=0}} \{<x_1,s_1> - \frac{1}{2} <s_1,s_1> \|x_o\|\}$$

has the value $(<x_1,x_1> - <s_o,x_1>^2)/2\|x_o\|$, which is achieved by $s_1 = (x_1 - <s_o,x_1> s_o)/\|x_o\|$. Hence we obtain

$$F_2 = \sup_{(s_o,s_1) \in S_1} \sup_{s_2 \in T[S;s_o,s_1]} F_2(s_o,s_1,s_2)$$

$$= \frac{1}{\|x_o\|} \left(<x_2,x_o> + \frac{1}{2} \left(<x_1,x_1> - \left(\frac{<x_o,x_1>}{\|x_o\|}\right)^2\right)\right).$$

Using the fact that $E(x)$ is a singleton for $x \neq \theta$ it is straightforward to check that hypothesis (H4) is satisfied. Thus we have proved that F_1 and F_2 are differentials of the Hilbertspace norm, and by Theorem 5 we obtain

<u>Theorem 10.</u> Let $M \subset H$ be a nonempty subset and let $y \in H$. Suppose that $x_o \in M$ is a local solution of the minimum norm problem

$$\underset{x}{\text{minimize}} \{\|x-y\| \mid x \in M\}$$

and that $\|x_o-y\| > 0$. Then

(24) $\quad <x_o-y,x_1> \geq 0 \qquad$ for all $x_1 \in T[M;x_o]$

and

(25) $\quad <x_o-y,x_2> + \frac{1}{2} \|x_1\|^2 \geq 0 \quad$ for all $x_2 \in T[M;x_o,x_1]$ and for all $x_1 \in T[M;x_o]$ with $<x_o-y,x_1> = 0$.

<u>Remarks.</u>

1. It should be noted that this result could also be proved by direct arguments very simple.

2. Using F_1 and F_2 and the same arguments in proving Theorem 8 it is easy to derive necessary conditions for best uniform approximations in the space C(s,H) of all vector-valued continuous functions on an arbitrary compact Hausdorff space S.

We now consider another important special case. We suppose that $(X,\|\cdot\|)$ is a normed linear space (with the norm topology $\mathcal{U}$) and that $(Z,<\cdot,\cdot>)$ is a Hilbert space (with the induced topology $\mathcal{W}$). In order to calculate F_1 and F_2 we shall make the following assumptions on $F : B \times R \to \mathbb{R}$ at x_o.

(A) $x_o \in \text{int } B$, $S_o := E(x_o) \subset \text{int } R$. For every $s_o \in S_o$ F is twice Frêchet-differentiable at (x_o,s_o).

(B) $S_o \subset \text{int } S$. For every $s_o \in S_o$ there is a constant $m > 0$ such that $F_{ss}(x_o,s_o)(z,z) \leq - m \|z\|^2$ for all $z \in Z$.

It follows at once from hypothesis (A) that

$$F_1(s_o,s_1) := F_x(x_o,s_o)(x_1) + F_s(x_o,s_o)(s_1) \quad \text{and}$$

$$F_2(s_o,s_1,s_2) := F_x(x_o,s_o)(x_2) + \tfrac{1}{2} F_{xx}(x_o,s_o)(x_1,x_1) + F_{xs}(x_o,s_o)(x_1,s_1) + F_s(x_o,s_o)(s_2) + \tfrac{1}{2} F_{ss}(x_o,s_o)(s_1,s_1)$$

are the $T_{\mathcal{U}\times\mathcal{W}}[B \times S]$-differentials of $F : B \times R \to \mathbb{R}$ at (x_o,s_o) for every $s_o \in S_o$, $s_1 \in T[S;s_o]$, $s_2 \in T[S;s_o,s_1]$ and for every $x_1,x_2 \in X$.

Since $F_s(x_o,s_o)(s_1)$ and $F_s(x_o,s_o)(s_2) + \frac{1}{2} F_{ss}(x_o,s_o)(s_1,s_1)$ are $T_{\mathcal{W}}[S]$-differentials of $F(x_o,\cdot) : R \to \mathbb{R}$ at s_o it follows by Theorem 5 that

$$F_s(x_o,s_o)(s_1) \leq 0 \qquad \text{for all } s_1 \in T[S;s_o]$$

$$F_s(x_o,s_o)(s_2) + \tfrac{1}{2} F_{ss}(x_o,s_o)(s_1,s_1) \leq 0 \text{ for all } s_2 \in T[S;s_o,s_1]$$

and for all $s_1 \in T[S;s_o]$ with $F_s(x_o,s_o)(s_1) = \theta$. In particular, $F_{ss}(x_o,s_o)$ is negative semidefinite on $T[S;s_o] \cap \ker F_s(x_o,s_o)$.

If $T(S;s_o)$ and $T(S;s_o,s_1)$ are nonempty we have

(26) $$F_1 = \max_{s_o \in S_o} F_x(x_o,s_o)(x_1) = \max_{s_o \in S_o} f_1(s_o)$$

$$F_2 \leq \sup_{(s_o,s_1) \in S_1} \{f_2(s_o) + F_{xs}(x_o,s_o)(x_1,s_1)\}, \qquad \text{where}$$

$f_1(s_o), f_2(s_o)$ denote the $T_{\mathcal{U}}[B]$-differentials of $F(\cdot,s_o) : B \to \mathbb{R}$ at x_o (compare Theorem 7).

If in addition hypothesis (B) is satisfied it follows that $Z = T(S;s_o) = T(S;,s_o,s_1) \neq \phi$, $F_s(x_o,s_o) = \theta$ and $S_1 = S_o \times Z$, hence

(27) $$F_2 = \sup_{s_o \in S_o} \{f_2(s_o) + \sup_{s_1 \in Z} [\tfrac{1}{2} F_{ss}(x_o,s_o)(s_1,s_1) + F_{xs}(x_o,s_o)(x_1,s_1)]\} .$$

Let $A(s_o):Z \to Z$ denote the continuous negative definite linear operator defined by $\langle Az,z\rangle = F_{ss}(x_o,s_o)(z,z)$ for all $z \in Z$ and let $b(s_o) \in Z$ denote the element defined by $\langle b(s_o),z\rangle = F_{xs}(x_o,s_o)(x_1,z)$ for all $z \in Z$. Then by assumption (B) $A(s_o)^{-1}$ exists and the quadratic programming problem in (27) has the unique solution $s_1 = - A(s_o)^{-1}b(s_o)$. Thus we have

(28) $$F_2 = \sup_{s_o \in S_o} \{f_2(s_o) - \frac{1}{2} \langle b(s_o),A(s_o)^{-1}b(s_o)\rangle\} .$$

Finally, when X and Z are finite dimensional, then straightforward arguments show that hypothesis (H4) holds. Hence, F_1 and F_2 are the T[B]-differentials of f at x_o.

In many applications hypothesis (B) is violated because there are elements $s_o \in S_o \setminus \text{int } S$. For this reason we replace hypothesis (B) by the following assumption (cf. [6],[7],[11]).

(C) Let $Z = \mathbb{R}^n, n \in \mathbb{N}$, and let $h^i : Z \to \mathbb{R}$ be mappings for $i \in I, |I| < \infty$. Let $S := \{s \in Z | h^i(s) \leq 0,\ i \in I\}$ be compact. The following holds for every $s_o \in S_o$:

(a) For every $i \in I$ h is twice Frêchet-differentiable at x_o. The vectors $h^i_s(s_o)$, $i \in I(s_o)$ are linearly independent, where $I(s_o) := \{i \in I | h^i(s_o) = 0\}$.

There is a $z \in Z$ such that $h^i_s(s_o)(z) < 0$ for all $i \in I(s_o)$. For every $s_1 \in T[S;s_o]$, which satisfies $h^i_s(s_o)(s_1)=0$, $i \in I(s_o)$, there exists a $z \in Z$ such that $h^i_s(s_o)(z) + \frac{1}{2} h^i_{ss}(s_o)(s_1,s_1) < 0$ for all $i \in I(s_o)$.

(b) There are Lagrange-multipliers $l_i > 0$, $i \in I(s_o)$ and a constant $m > 0$ such that the Lagrange function defined by $L(s) := F(x_o,s) - \sum_{i \in I(s_o)} l_i h^i(s)$ satisfies $L_s(s_o) = \theta$ and $L_{ss}(s_o)(z,z) \leq - m \|z\|^2$ for every $z \in N(s_o) := \{z \in Z | h^i_s(s_o)(z) = 0 \text{ for all } i \in I(s_o)\}$.

As a first consequence it follows by Theorem 3 and 4 that

$$\phi \neq T(S;s_o) \subset \overline{T}(S;s_o) = T[S;s_o] = \{z \in Z | h^i_s(s_o)(z) \leq 0,\ i \in I(s_o)\}$$

and

$$\phi \neq T(S;s_o,s_1) \subset \overline{T}(S;s_o,s_1) = T[S;s_o,s_1]$$
$$= \{z \in Z | h^i_s(s_o)(z) + \frac{1}{2} h^i_{ss}(s_o)(s_1,s_1) \leq 0,\ i \in I(s_o)\}$$

for every $s_1 \in T[S;s_o] \cap N(s_o) = N(s_o)$.

Hence, we conclude that

$$F_1 = \max_{s_o \in S_o} f_1(s_o)$$

and

$$F_2 = \sup_{s_o \in S_o} \{f_2(s_o) + \sup_{s_1 \in N(s_o)} [\tfrac{1}{2} L_{ss}(s_o)(s_1,s_1) + F^o_{xs}(x_1,s_1)]\},$$

since $\sup_{s_2 \in T[S;s_o,s_1]} F^o_s(s_2) = -\frac{1}{2} \sum_{i \in I(s_o)} l_i h^i_{ss}(s_o)(s_1,s_1)$.

If $P_{N(s_o)}$ denotes the orthogonal projection onto the subspace $N(s_o)$ we obtain (28), where

$$A(s_o) := P_{N(s_o)} L_{ss}(s_o)\big|_{N(s_o)} \quad \text{and} \quad b(s_o) := P_{N(s_o)} F_{xs}(x_o,s_o)(x_1).$$

Using similar arguments as [6] we can prove that (H4) holds, provided that X is finite dimensional.

Thus the following necessary conditions hold for semi-infinite programming problems.

Theorem 11. Let $X = \mathbb{R}^p$, $Z = \mathbb{R}^n$, $p,n \in \mathbb{N}$. Suppose that $F : B \times R \to \mathbb{R}$ satisfies either hypotheses (A), (B) or hypotheses (A), (C).

Let $M \subset X$ be a nonempty subset. Suppose that $x_o \in M$ is a local solution of the semi-infinite programming problem

$$\underset{x}{\text{minimize}} \ \{\max_{s \in S} F(x,s) \mid x \in M\} .$$

Then

$$\max_{s_o \in S_o} F_x(x_o,s_o)(x_1) \geq 0 \text{ for all } x_1 \in T[M;x_o]$$

and

$$\max_{s_o \in S_o} \{F_x(x_o,s_o)(x_2) + \tfrac{1}{2} F_{xx}(x_o,s_o)(x_1,x_1) + c(s_o,x_1)\} \geq 0$$

for all $x_2 \in T[M;x_o,x_1]$ and for all $x_1 \in T[M;x_o]$ with

$\max_{s_o \in S_o} F_x(x_o,s_o)(x_1) = 0$, where $c(s_o,x_1) := -\frac{1}{2} \langle b(s_o), A(s_o)^{-1} b(s_o)\rangle \geq 0$

is defined in the proof above.

Remarks.

1. It should be noted that S_o is a finite set in this case.

2. It is clear that the preceeding discussion applies also to semi-infinite programming problems of the type E4.

3. Hettich [6] considers a uniform approximation problem, where F is given by $F(x,s) := |G(x,s) - y(s)|$.

REFERENCES

1. DEM'YANOV, V. F. and MALOZEMOV, V. N., *Introduction to minimax*, John Wiley and Sons, New York, New York 1974.

2. DEM'YANOV, V. F., *On a minimax problem*, Soviet Math. Dokl., Vol. 10, pp. 828-832, 1969.

3. DEM'YANOV, V. F., *Second order directional derivatives of the maximum function*, Kibernetika Vol. 5, pp. 67-69, 1973.

4. DEM'YANOV, V. F., *The minimax problem with dependent constraints*, USSR Comp. Math. Phys., Vol. 12, pp. 299-307, 1972.

5. DEM'YANOV, V. F. and PEVNYĬ, A. B., *First and second order marginal values of mathematical programming problems*, Soviet Math. Dokl., Vol. 13, pp. 1502-1506, 1972.

6. HETTICH, R., *Kriterien zweiter Ordnung für lokal beste Approximationen*, Numer. Math., Vol. 22, pp. 409-417, 1974.

7. HETTICH, R., *Kriterien erster und zweiter Ordnung für lokal beste Approximationen bei Problemen mit Nebenbedingungen*, Numer. Math., Vol. 25, pp. 109-122, 1975.

8. HOFFMANN, K.-H. and KORNSTAEDT, H.-J., *Higher-Order Necessary Conditions in Abstract Mathematical Programming*, Journal of Optimization Theory and Applications, Vol. 26, pp. 531-566, 1978.

9. KORNSTAEDT, H.-J., *Notwendige Minimalbedingungen*, Technische Universität Berlin, Habilitationsschrift, 1978.

10. LUSTERNIK, L. A. and SOBOLEV, V. H., *Elements of Functional Analysis*, John Wiley and Sons, New York, New York, 1974.

11. WETTERLING, W., *Definitheitsbedingungen für relative Extrema bei Optimierungs- und Approximationsaufgaben*, Numer. Math., Vol 15, pp. 122-136, 1970.

ON NUMERICAL ANALYSIS IN SEMI-INFINITE PROGRAMMING

Sven-Åke Gustafson
Department of Numerical Analysis
Royal Institute of Technology
S-100 44 STOCKHOLM 70, Sweden

Abstract. In this paper which is a companion to [9] we shall discuss the theoretical questions which arise by the computational treatment of semi-infinite programs. Fairly strong regularity assumptions will be needed to insure satisfactory results of the computational schemes described here and which have proved effective in actual calculations.

1. Notations and preliminaries

We immediately introduce the concept of a dual pair of semi-infinite programs (SIP's). We will adhere to the notations in [3] and [9]. We will work with the entities:

S is a fixed but arbitrary index-set
$a_1, a_2, \dots, a_n$ and b are n+1 functions defined on S
$c \in R^n$ a fix vector.

With these data we now define the two problems:

Program (P). Minimize the linear form

(1a) $$c^T y$$

over all vectors $y \in R^n$ subject to the constraints

(1b) $$a(s)^T y \geq b(s), \quad s \in S \qquad //$$

and

Program (D). Determine the integer q, the subset $\{s_1, s_2, \dots, s_q\} \subset S$ and the reals $x_1, x_2, \dots, x_q$ such that the expression

(2a) $$\sum_{i=1}^{q} x_i b(s_i)$$

is maximized under the constraints

(2b) $$\sum_{i=1}^{q} x_i \, a(s_i) = c$$

(2c) $$x_i \geq 0, \quad i = 1,2,\dots,q \qquad //$$

Here $a(s) \in R^n$ is the vector whose components are $a_r(s)$, $r = 1,2,\dots,n$. In the applications it is very common that the index set S has the form

(3) $$S = \bigcup_{j=0}^{\ell} S_j$$

where S_0 has finitely many elements only. We reformulate Programs (P) and (D) for this case and get

Program (P). Minimize the linear form

(4a) $$c^T y$$

over all vectors $y \in R^n$ subject to the constraints

(4b) $$a(s)^T y \geq b(s), \quad s \in S_j, \quad j = 0,1,\dots,\ell$$ //

and

Program (D). Determine the integers q_j, $j = 0,1,\dots,\ell$ the subsets $\left\{s_1^j, s_2^j, \dots, s_{q_j}^j\right\} \subset S_j$, $j = 0,1,\dots,\ell$ and the reals $x_1^j, x_2^j, \dots, x_{q_j}^j$ such, that the expression

(5a) $$\sum_{j=0}^{\ell} \sum_{i=1}^{q_j} x_i^j \, b(s_i^j)$$

is rendered a maximum under the constraints

(5b) $$\sum_{j=0}^{\ell} \sum_{i=1}^{q_j} x_i^j \, a(s_i^j) = c$$

(5c) $$x_i^j \geq 0, \quad i = 1,2,\dots,q_j, \quad j = 0,1,\dots,\ell .$$ //

Obviously the problem (5) is a special case of (2). Note that we use in this paper the convention that when referring to an entire group of formulas whose labels consist of a certain number followed by letters we use the number only. Thus by (5) we mean (5a), (5b) and (5c).

In a computer we can only store finitely many numbers with a limited precision. Further we can only carry out finitely many arithmetic operations. Thus it must be possible to determine whether an arbitrary s belongs to S by performing only finitely many such operations. We shall normally assume that $S \subset R^k$, $k < \infty$ although some of the results will be formulated for a slightly more general situation. The functions a_r and b will be defined through computer programs which must be of finite length and call for only finitely many arithmetic operations.

An optimal solution of Program (P) will be represented by an ordered n-tuple. For Program (D) the situation is not that easy, since q may be arbitrarily large. However, the following result holds.

Lemma 1. Let Program (D) assume its optimal value. Then it has an optimal solution specified by q, $\{s_1, s_2, \dots, s_q\}$, $x_1, x_2, \dots, x_q$ such that

(6a) i) $q \leq n$

(6b) ii) $x_i > 0$, $i = 1,2,\dots,q$

(6c) iii) $a(s_i)$, $i = 1,2,\ldots,q$ are linearly independent.

The elementary *proof* can be carried out straightforwardly using properties of convexity and linear dependency. See e.g. [3] p 120-121.

We will only discuss the representation of an optimal solution of Program (D) satisfying (6) when $S \subset R^k$ and with S given by (3). The solution

$$q, \quad \{s_1,s_2,\ldots,s_q\} \quad x_1,x_2,\ldots,x_q$$

is specified by among other things q vectors $s_1,s_2,\ldots,s_q$ which can be ordered in q! different ways. In order to remove this ambiguity in the computer representation we introduce a canonical ordering in two phases as follows. First we order the vectors $s_1,s_2,\ldots,s_q$ such that first come the elements out of $\{s_1,s_2,\ldots,s_q\}$ belonging to S_0 (if any) then those in S_1, etc. Next we order the elements in each S_j in lexicographic order. (If $u \in R^k$ and $v \in R^k$, then u comes before v in the lexicographic order, if *either* $u_1 < v_1$ *or* there is an i, $2 \leq i \leq n$ such that $u_j = v_j$, $j = 1,2,\ldots,i-1$ and $u_i < v_i$). If we order $s_1,s_2,\ldots,s_q$ as described above than an optimal solution of Program (D) is uniquely specified by the integers $q_0,q_1,\ldots,q_\ell$, the vectors $s_1,s_2,\ldots,s_q$ and the reals $x_1,x_2,\ldots,x_q$, i.e. in total $1 + \ell + q(k+1)$ numbers.

Definition 1. Two dual pairs (P), (D) and ($\tilde{P}$), ($\tilde{D}$) specified by respectively S,a,b,c and S,$\tilde{a}$,$\tilde{b}$,$\tilde{c}$ are said to be *computationally equivalent*, if the computer representations coincide for the following pairs of numbers:

$$a_r(s) \text{ and } \tilde{a}_r(s), \quad r = 1,2,\ldots,n, \quad s \in S$$

$$b(s) \text{ and } \tilde{b}(s), \quad s \in S$$

$$c_r \text{ and } \tilde{c}_r, \quad r = 1,2,\ldots,n$$

//

Example 1. Put

$$b(s) = e^s \qquad \tilde{b}(s) = \sum_{k=0}^{20} \frac{s^k}{k!} \qquad -1 \leq s \leq 1$$

The representations of b and $\tilde{b}$ coincide for a computer working with the relative accuracy 10^{-8}. //

We now want to prove that if S is compact and a_r, $r = 1,2,\ldots,n$ and b continuous on S, then Programs (P) and (D) are computationally equivalent to linear programs. We first need

Definition 2. Let $T = \{t_1,t_2,\ldots,t_N\}$ be a finite subset of a set $S \subset R^k$. Then T will be called a *grid*. Denote by k(T) its convex hull. Next determine N continuous functions $\rho_1,\rho_2,\ldots,\rho_N$ such that

i) each $s \in k(T)$ has a representation

$$s = \sum_{j=1}^{N} \rho_j(s)t_j$$

ii) $\sum_{j=1}^{N} \rho_j(s) = 1$

iii) $\rho_j(s) \geq 0, \quad j = 1,2,\ldots,N$

iv) for each $s \in k(T)$ at most $k+1$ $\rho_j(s)$ are strictly positive.

(Note that ρ_j are not uniquely determined by i) through iv)).

Let now a function f be defined on T. Define Lf through

$$(Lf)(s) = \sum_{j=1}^{N} \rho_j(s) f(t_j)$$

Then L will be called the *positive linear interpolator* induced by T and $\rho_1, \rho_2, \ldots, \rho_N$. //

We next state

<u>Lemma 2</u>. Let S, T, k(T) and L be defined as in Definition 2. Then $y \in R^n$ satisfies the inequalities

$$a(t_j)^T y \geq b(t_j), \quad j = 1,2,\ldots,N$$

if and only if y satisfies the inequalities

$$La^T(s)y \geq Lb(s), \quad s \in k(T)$$

The proof is immediate. See e.g. [3] p 148. //

We now arrive at the result announced earlier.

<u>Theorem 1</u>. Let $S \subset R^k$ be a compact set, a_r, $r = 1,2,\ldots,n$ and b continuous functions on S. Then there is a finite subset $T \subset S$, $T = \{t_1, t_2, \ldots, t_N\}$ such that Program (P) is computationally equivalent to the task: Minimize the linear form

(7a) $$c^T y,$$

over all vectors $y \in R^n$ subject to the constraints

(7b) $$La(s)^T y \geq b(s), \quad s \in S$$

The problem (7) can be solved by means of linear programming. L is defined as in Definition 2.

<u>Proof</u>. Let $\varepsilon > 0$ be given. Since S is compact and a_r, $r = 1,2,\ldots,n$ and b are continuous on S it is possible to find a subset $T = \{t_1, t_2, \ldots, t_N\}$ and a corresponding positive linear interpolator L such that

i) $T \subset S \subset k(T)$

ii) $|La_r(s) - a_r(s)| < \varepsilon, \quad s \in S, \quad r = 1,2,\ldots,n$

iii) $|Lb(s) - b(s)| < \varepsilon, \quad s \in S$

L may be constructed e.g. through interpolating linearly in each of the k dimensions. By Lemma 2 (7b) is equivalent to

$$a(t_j)^T y \geq b(t_j), \quad j = 1,2,\dots,N$$

and hence Problem (7) is equivalent to a linear program with n variables and N constraints. By choosing ε small enough we achieve the desired result. //

As a consequence of Theorem 1 a SIP could be replaced by a linear program although the latter generally will be very large. This is the rationale behind the methods based on discretization where a SIP is approximated by a linear program. This can always be solved in a finite time. The iterative methods to be treated in Section 4 are often more rapid provided an initial approximation of the optimal solution y together with some other information are given. These methods have been discussed extensively in [12], [13]. However, for these methods convergence is not guaranteed in many practical situations. An initial approximation of the optimal solution may be constructed by means of linear programming and then refined by means of iterative schemes. In case of convergence failure the initial approximation is refined before iteration is recommenced. Such a combined approach is described in [3], [5], [10] and implemented in the computer codes in [2].

2. Wellposedness in SIP

If the optimal value v(P) of Program (P) varies very strongly by small perturbations in input data a, b, c then v(P) is not well determined by the computer representation of Program (P). We illustrate this by

Example 2. Minimize the linear form

$$y_1 + \frac{1}{2}y_2 + \left(\frac{1}{2} + \frac{\varepsilon}{3}\right)y_3$$

over all $y \in R^3$ subject to the constraints

$$y_1 + y_2 s + y_3(s + \varepsilon s^2) \geq e^s, \quad s \in [0,1]$$

We find: $v(p) = \frac{1}{2}(1+e) \quad \varepsilon = 0, \quad v(P) = \frac{1}{4}(3e^{1/3} + e), \quad \varepsilon \neq 0$ //

Example 3. Minimize the linear form

$$y_1 + (1+\varepsilon)y_2$$

over all $y \in R^2$ subject to the constraints

$$y_1 + y_2 s \geq -\frac{1}{1+s}, \quad s \in [0,1].$$

We get $v(P) = -\infty, \ \varepsilon > 0 \quad v(P) = -\frac{1}{2}, \ \varepsilon = 0 \quad v(P) = -\frac{1}{2+\varepsilon}, \ \varepsilon < 0$ //

In the two examples v(P) is poorly determined, if $|\varepsilon|$ is a small number in comparison to the computational errors.

We will now give conditions which insure that the problem to determine the value v(P) is well-posed. The corresponding optimal solution can, under the same conditions be shown to vary continuously with input data, provided it is unique. (In case of nonuniqueness we could take the optimal solution with the smallest Euclidean norm. Unfortunately this imposed uniqueness does not imply that the so selected solution varies continuously with input data, as easily seen from simple examples).

Lemma 3. Let (1b) be consistent and assume that c has a representation

$$\sum_{i=1}^{n} x_i a(s_i) = c \qquad x_i > 0, \quad s_i \in S, \quad i = 1,2,\ldots,n$$

and where $a(s_1), a(s_2), \ldots, a(s_n)$ are linearly independent. Then there is an $F > 0$ such that all optimal solutions y of Program (P) satisfy

$$|y_r| \le F, \quad r = 1,2,\ldots,n$$

Proof. By (7) on p 106 in [3] Program (P) has an optimal solution y. It must also met (1b). Hence there are numbers d_i such that

$$y^T a(s_i) = b(s_i) + d_i \text{ with } d_i \ge 0, \; i = 1,2,\ldots,n \tag{8}$$

We get also

$$y^T c = \sum_{i=1}^{n} x_i b(s_i) + \sum_{i=1}^{n} x_i d_i$$

Since $x_i > 0$ by assumption there is a $B > 0$ such that

$$0 \le d_i \le B, \quad i = 1,2,\ldots,n \tag{9}$$

We may look upon (8) as a linear system with $y \in R^n$ as the unknown. Since $a(s_1)$, $a(s_2)$, ..., $a(s_n)$ are assumed to be linearly independent this system has a nonsingular matrix. Due to (9) its right hand side is bounded by $B + \max_i |b(s_i)|$ and the desired result follows. //

We next derive an expression for the change in the value v(P) which is caused by perturbations in input data a, b and c.

Lemma 4. Let S be a compact set, a_r and $\tilde{a}_r$, $r = 1,2,\ldots,n$ and b, $\tilde{b}$ be continuous functions on S. c and $\tilde{c}$ are given vectors. With these data we form the two tasks:

Program (P): Minimize $c^T y$

subject to $\quad a(s)^T y \ge b(s), \quad s \in S$

Program ($\tilde{P}$): Minimize $\tilde{c}^T y$

subject to $\quad \tilde{a}(s)^T y \ge \tilde{b}(s), \quad s \in S$

We make the assumptions:

i) (P) and ($\tilde{P}$) have optimal solutions y and $\tilde{y}$ respectively

ii) There is a vector y^S and a number $\omega > 0$ such that

$$a(s)^T y^S - b(s) \ge \omega, \; s \in S \quad \tilde{a}(s)^T y^S - \tilde{b}(s) \ge \omega, \; s \in S$$

Introduce the notations:

F is the largest of the 2n numbers $|y_r|$ and $|\tilde{y}_r|$, $r = 1,2,\dots,n$,

$$\delta_r = \max_{s\in S}|a_r(s) - \tilde{a}_r(s)|, \quad r = 1,2,\dots,n, \quad \delta_{n+1} = \max_{s\in S}|b(s) - \tilde{b}(s)|$$

$$\Delta = \delta_{n+1} + F\sum_{r=1}^{n}\delta_r$$

$$C_r = \max(|c_r|,|\tilde{c}_r|), \quad r = 1,2,\dots,n$$

$$Y_r = \max(|y_r^S - y_r|, \ |y_r^S - \tilde{y}_r|), \quad r = 1,2,\dots,n$$

Then we may state

$$|v(P) - v(\tilde{P})| \le F\sum_{r=1}^{n}|\tilde{c}_r - c_r| + \frac{\Delta}{\omega+\Delta}\sum_{r=1}^{n}C_rY_r$$

Proof. We want to construct a vector $\bar{y}$ of the form $\bar{y} = (1-\lambda)y + \lambda y^S = y + \lambda(y^S-y)$ which is feasible for $\tilde{P}$. We get first

$$\tilde{a}(s)^Ty - \tilde{b}(s) = a(s)^Ty - b(s) + \{\tilde{a}(s)^T - a(s)^T\}y + b(s) - \tilde{b}(s)$$

Since y is feasible for (P), $a(s)^Ty - b(s) \ge 0$.

Thus

$$\tilde{a}(s)^Ty - \tilde{b}(s) \ge -(\delta_{n+1} + F\sum_{r=1}^{n}\delta_r) = -\Delta$$

Next

$$\tilde{a}(s)^T\bar{y} - \tilde{b}(s) = (1-\lambda)[\tilde{a}(s)^T\bar{y} - \tilde{b}(s)] + \lambda[\tilde{a}(s)^Ty^S - \tilde{b}(s)] \ge -(1-\lambda)\Delta + \lambda\omega = 0$$

$$\text{for } \lambda = \Delta/(\omega+\Delta)^{-1}$$

With this choice of λ, $\bar{y}$ is hence feasible for $\tilde{P}$ and we get therefore

$$v(\tilde{P}) \le \tilde{c}^T\bar{y} = \tilde{c}^T\{y + \lambda(y^S-y)\} = c^Ty + (\tilde{c}-c)^Ty + \lambda\tilde{c}^T(y^S-y)$$

Since $v(P) = c^Ty$ we obtain

$$v(\tilde{P}) - v(P) \le (\tilde{c}-c)^Ty + \lambda\tilde{c}^T(y^S-y)$$

Using the notations introduced above we arrive at

$$(10) \qquad v(\tilde{P}) - v(P) \le F\sum_{r=1}^{n}|\tilde{c}_r - c_r| + \frac{\Delta}{\omega+\Delta}\sum_{r=1}^{n}C_r\,Y_r$$

If we interchange the roles of P and $\tilde{P}$ in the calculation above we establish that $v(P) - v(\tilde{P})$ is also bounded by the right hand side of (10) and hence the assertion of the lemma is valid. //

Definition 3. Let S be a set, a_r, $r = 1,2,\dots,n$ and b functions defined on S. If there is a vector y^S and a number $\omega > 0$ such that

$$(11) \qquad a(s)^Ty^S - b(s) \ge \omega, \quad s \in S,$$

then $a_1,a_2,\dots,a_n$ and b are said to meet *Slater's condition* over S. //

Definition 4. Let S be a compact set and let a_r, $r = 1,2,\ldots,n$ and b belong to C(S), the linear space of continuous functions defined on S and equipped with the maximum norm. Let $c \in R^n$ and define the linear space U of ordered triples $u = (a,b,c)$. Define $\|u\|$ through $\|u\| = \max(\|a_1\|, \|a_2\|,\ldots, \|a_n\|, \|b\|, |c_1|, |c_2|,\ldots, |c_n|)$. Then U is called the *dataspace* of Program (P). //

We now establish the main result of this Section:

Theorem 2. Let $u = (a,b,c)$ be as in Definition 4. Make the following assumptions

i) S is a compact subset of R^k

ii) $a_1,a_2,\ldots,a_n$ and b meet Slater's condition over S

iii) c has the representation required in Lemma 3.

On the dataspace U we introduce the norm given in Definition 4. Then we assert: There is a neighbourhood N(u) of u with the properties

a) The optimal value $v(\tilde{P})$ of each Program $\tilde{P}$ formed with respect to $\tilde{u} = (\tilde{a},\tilde{b},\tilde{c}|$ is a continuous function on N(u).

b) Provided each Program $\tilde{P}$ has a unique optimal solution $\tilde{y}$ it varies continuously as a function of input data $\tilde{a},\tilde{b},\tilde{c}$.

Proof. It is clear that there is a neighbourhood of u such that the assumptions ii) and iii) hold. Hence by the results on p 115 in [3] the corresponding Programs have optimal solutions. By a slight modification of the argument in the proof of Lemma 3 we established that these optimal solutions must be confined to a bounded subset of R^n. The statement a) is now an easy consequence of Lemma 4. Assume now that each Program $\tilde{P}$ has a unique optimal solution $\tilde{y}$. We want to establish that $\tilde{y}$ is a continuous function of $\tilde{u}$ at $\tilde{u} = u$. Assume the contrary. Let y^0 be the optimal solution corresponding to u. Thus there is a sequence $u^1, u^2, \ldots$ such that $u^\ell \to u$ but the corresponding optimal solutions y^ℓ do not converge to y^0. However, $\{y^\ell\}_0^\infty$ is confined to a bounded subset of R^n and hence has an accumulation point z. We thus select a subsequence $y^{\ell_k} \to z$. Since $u^{\ell_k} \to u$ we find that z is the unique optimal solution corresponding to u, thus $z = y^0$. We find that all convergent subsequences of $\{y^\ell\}$ have y^0 as limitpoint and this fact establishes the contradiction sought. //

3. Discrete approximations of semi-infinite programs

Let the assumptions of Theorem 2 prevail! Form a sequence of grids $\{T^\ell\}_1^\infty$ with

$$T^1 \subset T^2 \subset \ldots \subset S$$

and such

$$\lim_{\ell\to\infty} \max_{s\in S} \min_{t\in T^\ell} \|s-t\| \to 0$$

where $\| \ \|$ is a norm on R^k. Then there is a ℓ_0 such that for $\ell \geq \ell_0$ the linear program obtained when T^ℓ replaces S has an optimal solution. Assume now that the optimal solution y^ℓ is unique. By Theorem 2 the sequence $\{y^\ell\}_{\ell_0}^{\infty}$ must converge to an optimal solution of Program (P).

A similar result cannot be established for Program (D), however. We illustrate with

Example 4. Determine the integer q, the subset $\{s_1,s_2,\ldots,s_q\} \subset [0,1]$ and the reals $x_1,x_2,\ldots,x_q$ such that the expression

$$-\sum_{i=1}^{q} x_i/(1+s_i)$$

is maximized under the constraints

$$\sum_{i=1}^{q} x_i \, s_i^{r-1} = 1/r, \quad r = 1,2$$

$$\text{and } x_i \geq 0, \quad i = 1,2,\ldots,q$$

The optimal solution is $q = 1$ $x_1 = 1$ $s_1 = 1/2$ and the corresponding value is -2/3. Let us now discretize this problem and replace [0,1] with the grids $T^\ell = \{s_1^\ell, s_2^\ell, \ldots, s_\ell^\ell\}$ where

$$s_i^\ell = \frac{i-1}{\ell-1}, \quad i = 1,2,\ldots,\ell, \quad \ell = 2,3,\ldots$$

In the cases ℓ odd and ℓ even we find the unique optimal solutions:

$$\ell = 2m+1: \quad q = 1, \quad s_1^\ell = \frac{1}{2}, \quad x_1^\ell = 1 \quad \text{value} = -2/3$$

$$\ell = 2m: \quad q = 2, \quad s_1^\ell = \frac{m-1}{2m-1}, \ s_2^\ell = \frac{m}{2m-1}, \ x_1^\ell = x_2^\ell = \frac{1}{2}, \ \text{value} = -\frac{3}{2}\,\frac{(2m-1)^2}{(3m-2)(3m-1)}.$$

//

Large values of ℓ correspond to small perturbations in input data. We may even by selecting ℓ *odd* get an arbitrarily close approximation where the optimal solution of the discretized problem has $q = 2$ but the originally SIP has $q = 1$. For further illustrations of this phenomena see e.g. [2], [3], [8]. Thus the problem to calculate q is not well posed. However, we can derive the following results.

Lemma 5. Let Program D have an optimal solution q, $\{s_1,s_2,\ldots,s_q\}$, $x_1,x_2,\ldots,x_q$ corresponding to the value v(D). Let further $a_1,a_2,\ldots,a_n$ and b meet Slater's condition over S. Then we can state, using the notations of Definition 3.

$$\sum_{i=1}^{q} x_i \leq \omega^{-1}[c^T y^S - v(D)] \tag{12}$$

Proof. Since q, $\{s_1,s_2,\ldots,s_q\}$ and $x_1,x_2,\ldots,x_q$ is also feasible for Program (D) we have

$$c^T y^S - v(D) = \sum_{i=1}^{q} x_i \{a(s_i)^T y^S - b(s_i)\} \geq \omega \sum_{i=1}^{q} x_i \quad \text{giving (12).}$$

//

We next introduce

Definition 5. The functions $a_1, a_2, \ldots, a_n$ which are defined on a set S are said to meet *Krein's condition* over S, if there is a vector $y^K \in R^n$ and a number $\lambda > 0$ such that

$$(13) \qquad a(s)^T y^K \geq \lambda, \quad s \in S \qquad //$$

We note that if Krein's condition is met by the continuous functions $a_1, a_2, \ldots, a_n$ over a compact set S, then the n+1 functions $a_1, a_2, \ldots, a_n$ and b meet Slater's condition over S for all continuous b. In particular, Krein's condition is met if $a_1(s) = 1$, a case often occurring in practice.

Lemma 6. Let q, $\{s_1, s_2, \ldots, s_q\} \subset S$ and $x_1, x_2, \ldots, x_q$ be feasible for Program (D). If $a_1, a_2, \ldots, a_n$ meet Krein's condition over S then, using the notations of Definition 5

$$(14) \qquad \sum_{i=1}^{q} x_i \leq \lambda^{-1} c^T y^K$$

Proof. Since q, $\{s_1, s_2, \ldots, s_q\}$ and $x_1, x_2, \ldots, x_q$ is feasible for Program (D),

$$c^T y^K = \sum_{i=1}^{q} x_i \; a(s_i)^T y^K \geq \lambda \sum_{i=1}^{q} x_i$$

and hence (14) follows. //

One could say that Lemmas 5 and 6 are the counterpart of Lemma 3. If all of these Lemmas apply, they should guarantee that numbers of moderate size only occur during the numerical treatment of an SIP. If very large numbers appear nevertheless this fact signals numerical difficulties. One source of these troubles could be that $a_1, a_2, \ldots, a_n, b$ are almost linearly dependent, i.e. one or more of the a_r:s can be approximated accurately with a linear combination of the remaining ones: Then the corresponding component in (2b) should be dropped and n decreased by 1. In particular, if b is approximated accurately by a linear combination of $a_1, a_2, \ldots, a_n$, then v(D) is correspondingly well determined by the conditions (2b), (2c), and the task to evaluate v(D) should not be treated as a SIP. In the general situation one could contemplate replacing $a_1, a_2, \ldots, a_n, b$ by linear combinations of the same functions selected to be "more linearly independent". This could be achieved using a scheme similar to the modified Gram-Schmidt orthogonalization procedure.

We next discuss the solution of a discretized SIP by means of the simplex method. It is essential that stable updating of the inverse is performed e.g. according to [3] Kap IV.

Let $\{s_1, s_2, \ldots, s_n\} \subset S$ specify a basic solution obtained by the simplex method. Then $a(s_1), a(s_2), \ldots, a(s_n)$ are linearly independent and

$$(15a) \qquad \sum_{i=1}^{n} x_i a(s_i) = c \quad \text{with} \quad x_i \geq 0, \quad i = 1,2,\ldots,n$$

To test for optimality we determine $y \in R^n$ from the system

(15b) $$y^T a(s_i) = b(s_i) \quad i = 1,2,\dots,n$$

Next we determine s* as the point minimizing $y^T a(s) - b(s)$ on the grid used for discretization. If the corresponding value is nonnegative, y is an optimal solution. Otherwise another simplex step must be performed and s* must replace one of the s_i:s in (15a). To determine *which* we need to solve the system

(15c) $$\sum_{i=1}^{n} \rho_i a(s_i) = a(s^*)$$

All of the systems (15) have a regular matrix and hence unique solutions. But their condition can be poor, if the grid is fine and some of the s_i lie closely together since then the vectors $a(s_i)$ are almost linearly dependent. Since the analysis of the general situation is complicated but does not give more insight we discuss only the special case $n = 2\ell$ and s_{2i}, s_{2i-1} lying closely together for $i = 1,2,\dots,\ell$. We assume also that S is an interval. a_r, $r = 1,2,\dots,n$ and b are also assumed to be continuously differentiable. (15a) takes the form

$$\sum_{i=1}^{\ell} \{x_{2i-1}\, a(s_{2i-1}) + x_{2i}\, a(s_{2i})\} = c$$

or

(16a) $$\sum_{i=1}^{\ell} \left\{u_i a(s_{2i-1}) + v_i \frac{a(s_{2i})-a(s_{2i-1})}{s_{2i}-s_{2i-1}}\right\} = c$$

where $u_i = x_{2i-1} + x_{2i}$ $v_i = x_{2i}(s_{2i}-s_{2i-1})$, $i = 1,2,\dots,\ell$. We see from (16a) that u_i and v_i are the solution of a system whose condition number remains bounded when $s_{2i} - s_{2i-1}$ becomes small. This means that the *sums* $x_{2i-1} + x_{2i}$ remain well determined provided that the quotients $(a(s_{2i}) - a(s_{2i-1}))/(s_{2i} - s_{2i-1})$ can be evaluated with full accuracy, e.g. if the derivative of a_r is numerically available. But the individual values x_{2i}, x_{2i-1} are obtained from the solution of a system whose inverse is roughly proportional to $\max_i (s_{2i} - s_{2i-1})^{-1}$. If only the functional values $a_r(s)$ are available, then the condition of the problem to determine u_i and v_i is also proportional to $(s_{2i} - s_{2i-1})^{-1}$ as apparent from (16a). These results are independent of ℓ, the number of groups s_{2i-1}, s_{2i} in (15). For (15c) analogous statements can be made. We turn now to (15b) which may be written

(16b) $$\sum_{r=1}^{n} y_r a_r(s_{2i}) = b(s_{2i}), \quad \sum_{r=1}^{n} y_r \frac{a_r(s_{2i})-a_r(s_{2i-1})}{s_{2i}-s_{2i-1}} = \frac{b(s_{2i})-b(s_{2i-1})}{s_{2i}-s_{2i-1}}$$

Thus the condition of the problem to determine y is roughly independent of $s_1,s_2,\dots,s_{2\ell}$ provided that the quotients of the form $(a(s_{2i})-a(s_{2i-1}))/(s_{2i}-s_{2i-1})$ can be evaluated with full accuracy e.g. by means of derivatives of a_r. Otherwise y is solution of a system of equations whose condition number is approximately proportional to $\max_i (s_{2i}-s_{2i-1})^{-1}$. In this discussion we have tacitly assumed that the systems, which arise from (16a), (16b) when the quotients are replaced by the corresponding derivatives have nonzero determinants. The analysis carried out here may be extended to the case when $S \subset R^k$ $k > 1$. See also [8].

The optimal solutions of the discretized versions of a SIP generally have $q = n$ even if the optimal solutions of the SIP itself have $q < n$. From the discussion above of the condition of the systems (15) we conclude that in this case there is an optimal roughness of the grid when the discretized version gives the best estimate of the optimal solution of Program (P). The decrease of the discretization errors which is obtained must namely be balanced off against the deterioration of the condition of the systems (15), which results from making the grid finer. But if $q = n$ the condition of the system should be largely independent of the fineness of the grid. Lastly if derivatives of a_r, $r = 1,2,\dots,n$ and b are available this information could be used to replace (15) with (16) and hence achieve that the condition of the systems becomes largely unaffected by the fineness of the grid even if $q < n$. However, the derivatives could profitably be used in the iterative schemes to be treated in the next Section.

4. Iterative schemes

Let q, $\{s_1, s_2, \dots, s_q\}$ and $x_1, x_2, \dots, x_q$ be an optimal solution of Program (D), $y \in R^n$ an optimal solution of Program (P). Then these entities must simultaneously satisfy (1b), (2b) and (2c). We have also the complementary slackness equation (see e.g. [3], [5], [10]).

(17) $$x_i(y^T a(s_i) - b(s_i)) = 0, \quad i = 1,2,\dots,q$$

Put

(18) $$f(s) = y^T a(s) - b(s)$$

By (1b) we have $f(s) \geq 0$, $s \in S$ and by (17) $f(s_i) = 0$, $i = 1,2,\dots,n$. Therefore we conclude:

(19) $$f \text{ has a global minimum at } x_i, \text{ if } x_i > 0, \; i = 1,2,\dots,q$$

This fact is used to generate further relations. Thus if f is differentiable and s_i is in the interior of S, then we get

$$\nabla f(s_i) = 0$$

since the *global* minimum at s_i is also a *local* minimum. But if s_i is at the boundary of S, then we combine the equations arising from the fact that s_i is at the boundary with the circumstance that a local minimum for f occurs there. In [12] this is expressed in a uniform manner by means of so-called Kuhn-Tucker conditions. Then S is defined as the requirement that certain inequalities must be satisfied. In all problems studied so far (19) generates k independent equations, if $S \subset R^k$.

To summarize, the optimality conditions may be written as follows

(20a) $$\sum_{i=1}^{q} x_i a(s_i) = c$$

(20b) $$x_i f(s_i) = 0, \ i = 1,2,\ldots,q$$

(20c) f has a local minimum at x_i, $i = 1,2,\ldots,q$, where f is defined by (18)

(21a) $$x_i \geq 0, \quad i = 1,2,\ldots,q$$

(21b) $$f(s) \geq 0 \quad s \in S$$

The relations (20) are combined into a nonlinear system of equations from which optimal solutions of Programs (P) and (D) are calculated. It must then be verified that the proposed solutions meet (21). Note that (21b) generally defines infinitely many inequalities and hence cannot be checked computationally by means of finitely many operations. Sometimes analytical methods can be used to confirm that (21b) is met for a calculated optimal solutions. In many practical applications one may only be able to verify the statement that

$$f(s) \geq -\delta, \quad s \in S$$

where δ is a known positive number. Then it is not certain that the calculated vector $\bar{y}$ is optimal for Program (D). Bounds for the errors in v(P) caused by substituting y for the (unknown) optimal solution are given in [10].

We next describe how to construct the system (20) and an approximate solution to it from an optimal solution $\tilde{q}$, $\{\tilde{s}_1,\tilde{s}_2,\ldots,\tilde{s}_q\}$, $\tilde{x}_1,\tilde{x}_2,\ldots,\tilde{x}_{\tilde{q}}$, $\tilde{y}$ of a discretized version of Programs (P) and (D) corresponding to a certain grid T. Define $\tilde{f}$ through

$$\tilde{f}(s) = a(s)^T\tilde{y} - b(s), \quad s \in S$$

Then $\tilde{f}(s) \geq 0$, $s \in T$ but $\tilde{f}(s) < 0$ is possible for $s \notin T$.

We take now q equal to the number of local minima of f which are such that the corresponding values of f are nonpositive. Denote the positions of these minima by $\sigma_1,\sigma_2,\ldots,\sigma_q$. With each σ_i we associate the $\tilde{s}_j$:s lying closest to σ_i. Let G_i be the set of the corresponding j-values. The vectors $\tilde{s}_j$ with j belonging to a certain G_i generally lie closely together. As approximation for x_i we take the sum of those $\tilde{x}_j$:s with j in G_i. As approximation for s_i we could take either the center of gravity of the $\tilde{s}_j$:s with $j \in G_i$ and weighted with the x_j:s or we could use σ_i itself. The former strategy tends to give a slightly smaller residuals in (20a) while the latter on the other hand should give zero residuals in the equations obtained from (20c). The calculated vector $\tilde{y}$ is taken as an initial estimate for y.

A general approach is now to linearize the system formed from (20) and improve the approximate solution by means of Newton-Raphson's method. Thus in each step we must solve a linear system with $n+(k+1)q$ equations. This is the method implemented in the computer codes of [2] where suitable stopping rules are given. Alternative methods are presented in [12], where rates of convergence are discussed in a general context. Sometimes (20) has a special structure which may be exploited to save labour. We mention two such instances:

a) $q=n$. Then the equations generated by (20b), (20c) can be solved independently of those in (20a). Furthermore, in each iteration step one may first use (20b) to improve the estimate of y then (20c) to correct the estimates of s_i.

b) (20a) can be solved independently. This occurs e.g. when S is a subset of the real line and $a_1, a_2, \ldots, a_n$ and b form a Čebyšev system over S. Then the optimal solutions of Program (D) and the problem obtained when we *minimize* (2a) subject to (2b), (2c) correspond to generalized quadrature rules of the Gaussian type. A numerical method working for general Čebyšev systems is given in [6] but the very important special case when $a_r(s) = s^{r-1}$ is more efficiently solved by means of the method described in [4].

However, even in the general case a certain simplification is possible. Upon linearizing (20) we may use (20c) to express the corrections of s_i in terms of those of y. This relation is substituted into the linear equations obtained from (20a), (20b) whereupon the corrections in y and x_i are calculated. Then the improvements in s_i are found from (20c). Thus a decomposition of the linear system which must be solved in each Newton-Raphson iteration is achieved. Related ideas are discussed in [13].

In [11] it is shown that provided certain regularity conditions are met, Newton-Raphson's method converges to an approximate solution with small residuals even if the error in the solution itself may be relatively large. From Lemma 4 it transpires, that then the associated error in the optimal value is small since the residuals in (20a) correspond to $\tilde{c}_r - c_r$ and those of (20b) govern the size of Δ. Remaining numbers in the formula of the assertion of Lemma 4 are of moderate size when the assumptions of Lemmas 3 and 5 are valid.

The favourable theoretical results what regards the stability of computational schemes for solving SIP's agree with the experiences gained from actual calculations.

REFERENCES

[1] Charnes, A., W.W. Cooper and K.O. Kortanek, *Duality, Haar programs and finite sequence spaces*, Proc. Nat. Acad. Sci., U.S. 48 (1962), 783-786

[2] Fahlander, K., *Computer programs for semi-infinite optimization*, TRITA-NA-7312, Dept of Numerical Analysis, Royal Institute of Technology, S-10044 Stockholm 70, Sweden, 1973

[3] Glashoff, K. and S.-Å. Gustafson, *Einführung in die Lineare Optimierung*, Wissenschaftliche Buchgesellschaft, Darmstadt, 1978

[4] Golub, G.H. and J.H. Welsch, *Calculation of Gauss quadrature rules*, Math. Comp. 23 (1969), 221-230

[5] Gustafson, S.-Å., *On the computational solution of a class of generalized moment problems*, SIAM J Numer. Anal. 7 (1970), 343-357

[6] Gustafson, S.-Å., *Die Berechnung von verallgemeinerten Quadraturformeln vom Gausschen Typus, eine Optimierungsaufgabe*, in L. Collatz, W. Wetterling (Eds): Numerische Methoden bei Optimierungsaufgaben, ISNM 17 (1973), 59-71, Birkhäuser, Basel

[7] Gustafson, S.-Å., *On computational applications of the theory of moment problems*, Rocky Mountain J on Math 4 (1974), 227-240

[8] Gustafson, S.-Å., *Stability aspects on the numerical treatment of linear semi-infinite programs*, TRITA-NA-7604, Dept of Numerical Analysis, Royal Institute of Technology, S-10044 Stockholm 70, Sweden, 1976

[9] Gustafson, S.-Å., *On semi-infinite programming in numerical analysis*, this volume

[10] Gustafson, S.-Å. and K.O. Kortanek, *Numerical treatment of a class of semi-infinite programming problems*, NRLQ 20 (1973), 477-504

[11] Gustafson, S.-Å., G. Dahlquist and L. Reichel, *Remarks on the convergence of Newton-Raphson's method in the presence of round-offs*, TRITA-NA-7615, Dept of Numerical Analysis, Royal Institute of Technology, S-10044 Stockholm 70, Sweden

[12] Hettich, R., *Kriterien zweiter Ordnung für lokal beste Approximation*, Numer. Math. 22 (1974), 409-417

[13] Hettich, R., *A Newton-method for nonlinear Chebyshev approximation* in R. Schaback, K. Scherer (Eds): Approximation Theory, Lecture Notes in Math 556 (1976), 222-236, Springer, Berlin-Heidelberg, New York

A CENTRAL-CUTTING-PLANE ALGORITHM FOR SEMI-INFINITE PROGRAMMING PROBLEMS

P. R. Gribik
Department of Mathematics
Carnegie-Mellon University
Pittsburgh, PA 15213 USA

Abstract: We give a cutting-plane algorithm for a class of linear semi-infinite programming problems that is based on the central-cutting-plane algorithm of Elzinga and Moore for convex programs. This algorithm has the property of being able to generate a cut from any violated or tight constraint without affecting convergence or convergence rate of the algorithm. Thus, to insure convergence, one is not required to find the most violated (or nearly most violated) constraint as required by the alternating algorithm.

1. Introduction

In this paper, we will consider a cutting-plane solution technique for semi-infinite programming problems of the following form:

Program D:

$$\text{Find } V_D = \min c^T x$$
$$\text{for all } x \in \mathbb{R}^n \text{ subject to}$$
$$u^T(t)x \geq \lambda(t) \quad \text{for all } t \in S$$
$$x \in H.$$

We will make the following assumptions about the data in Program D:

(i) $c \in \mathbb{R}^n$, $c \neq 0$;

(ii) S is a compact subset of $\mathbb{R}^m$;

(iii) H is a compact convex subset of $\mathbb{R}^n$;

(iv) $u(\cdot) : S \to \mathbb{R}^n$ is continuous on S;

(v) $\lambda(\cdot) : S \to \mathbb{R}$ is continuous on S;

(vi) there is an $\hat{x} \in H$ for which

$$u^T(t)\hat{x} > \lambda(t) \quad \text{for all } t \in S$$

and which is not optimal for Program D.

One of the first cutting-plane methods for solving semi-infinite programs is the alternating algorithm of Gustafson and Kortanek [13]. This technique is based on the principles of the cutting-plane method of Kelley [15] for convex programs. For Program D, the alternating algorithm has the following form.

Alternating Algorithm for Program D:

Step 0: Let D_o be the program

$$\min \; c^T x$$
$$\text{for all } x \in \mathbb{R}^n \text{ subject to}$$
$$x \in H.$$

Let $k = 1$.

Step 1: Let $x_k \in \mathbb{R}^n$ be a solution to

$$\text{Program } D_{k-1}.$$

Step 2: Solve

$$\min_{t \in S} (u^T(t)x_k - \lambda(t)).$$

Let t_k be an element in S at which the minimum is attained. If

$$u^T(t_k)x_k - \lambda(t_k) \geq 0,$$

stop; x_k is optimal for Program D.

Step 3: Form Program D_k by adding the constraint

$$u^T(t_k)x \geq \lambda(t_k)$$

to Program D_{k-1}. Set $k = k + 1$ and go to Step 1.

In [13], Gustafson and Kortanek show that the limit points of the sequence $\{x_k\}_{k=1}^{\infty}$ are optimal for Program D if the method does not terminate.

If H is polyhedral, each of the programs D_k is a linear program. In this case, Step 1 is not difficult to perform. However, in Step 2, we must solve

$$\min_{t \in S} u^T(t)x_k - \lambda(t).$$

In general, we would expect this to be a difficult mathematical programming problem.

If x_k is infeasible, Step 2 requires that we find the most violated constraint. Gustafson and Kortanek recognized the difficulty of this problem, and in [14] modified the algorithm so that one must find a "sufficiently" violated constraint. Their modification consisted of replacing Step 0 and Step 2 by the following:

Step 0^*: Form Program D_o as in Step 0. Choose a sequence $\{\epsilon_k\}_{k=1}^{\infty}$ of non-negative numbers such that

$$\lim_{k\to\infty} \epsilon_k \to 0.$$

Step 2^*: Let

$$\delta(x_k) = \min_{t\in S} (u^T(t)x_k - \lambda(t)).$$

If $\delta(x_k) \geq 0$, stop; x_k is optimal for Program D. Otherwise seek a $t_k \in S$ which satisfies

$$u^T(t)x_k - \lambda(t) < 0$$
$$u^T(t)x_k - \lambda(t) \leq \delta(x_k) + \epsilon_k.$$

While Step 2^* is easier to implement than Step 2, it is still difficult. When x_k is infeasible, it requires that one find a violated constraint that is "close" to being most violated; and as k increases, it must get "closer" and "closer" to being most violated.

In the next section a method is presented that is based on the central-cutting-plane algorithm of Elzinga and Moore [3] for convex programming problems. This method generates a sequence of feasible and infeasible points for Program D. If we are at an infeasible point, the method can generate a cut from any violated or tight constraint without affecting convergence. Thus we no longer must find a constraint close to being most violated if we are at an infeasible point. However, we must be able to distinguish between feasible and infeasible points. This too can be difficult, but it should be easier than finding the most violated constraint at each step. Also, there may be classes of problems for which determining whether a given point is feasible or infeasible is easy. One such problem from optimal experimental design is given in [8]. For this problem, performing the feasibility check is simple and finding a violated constraint for an infeasible point is easy, while finding the most violated constraint is difficult.

2. The Central-Cutting-Plane Method and Convergence

We will immediately present the algorithm. For $x \in \mathbb{R}^n$, let $\|x\|$ be the Euclidean norm, that is

$$\|x\| = (x^T x)^{1/2}.$$

Central-Cutting-Plane Method for Program D:

Step 0: Let $\bar{b}$ be strictly greater than V_D. Choose a constant β in (0,1). Let SD_o be the program

$$\begin{aligned} &\underline{\max}\ \sigma \\ &\underline{\text{subject to}} \\ &c^T x + \|c\|\sigma \le \bar{b} \\ &x \in H. \end{aligned}$$

Choose $y_o \in H$. Let $k = 1$.

Step 1: Let $(x_k, \sigma_k) \in \mathbb{R}^n \times \mathbb{R}$ be a solution to SD_{k-1}. If $\sigma_k = 0$, stop. Otherwise go to Step 2.

Step 2: Delete constraints from SD_{k-1} according to either or both of the deletion rules or do not delete constraints. Call the resulting problem SD_{k-1}.

Step 3: (i) If x_k is feasible for Program D, that is,

$$u^T(t)x_k \ge \lambda(t) \quad \text{for all} \quad t \in S, \tag{1}$$

add the constraint

$$c^T x + \|c\|\sigma \le c^T x_k$$

to Program SD_{k-1}. Set $y_k = x_k$.

(ii) Otherwise, find a $t_k \in S$ such that

$$u^T(t_k)x_k - \lambda(t_k) \le 0. \tag{2}$$

Add the constraint

$$u^T(t_k)x - \|u(t_k)\|\sigma \ge \lambda(t_k)$$

to Program SD_{k-1}. Set $y_k = y_{k-1}$.

In either case, call the resulting program SD_k. Set $k = k + 1$ and return to Step 1.

Deletion Rule 1: Delete the constraint $c^T x + \|c\|\sigma \le \bar{b}$ or any constraint generated by Step 3(i) in a previous iteration if x_k is feasible for Program D, that is,

$$\min_{t \in S} (u^T(t)x_k - \lambda(t)) \ge 0.$$

Deletion Rule 2: Delete a constraint from SD_{k-1} if:

(a) the constraint was generated by Step 3(ii) at the j-th iteration where $j < k$;

(b) $\sigma_k \le \beta\sigma_j$;

(c) the constraint was not tight in SD_{k-1} at (x_k,σ_k), that is,

$$u^T(t_j)x_k - \|u(t_j)\|\sigma_k > \lambda(t_j).$$

Suppose the algorithm does not terminate. Then a sequence of feasible points $\{y_k\}_{k=\hat{k}}^{\infty}$ is generated and the limit points of this sequence are optimal for Program D. The proof of this fact closely parallels the convergence proof given by Elzinga and Moore for their algorithm for convex programs. The remainder of this section will be devoted to this proof.

Lemma 1. Using either, both or neither of the deletion rules, if the algorithm does not terminate, $\lim_{k\to\infty} \sigma_k = 0$.

Proof: Since the set of feasible points for Program D is compact and not empty, there exists a point that is optimal for Program D. Let $x^*\in\mathbb{R}^n$ be optimal for Program D, then $(x^*,0)$ is obviously feasible for SD_k for all k. Hence, $\sigma_k \geq 0$ for all k. Deletion Rule 1 only drops constraints that are not binding. Therefore $\sigma_k \geq \sigma_{k+1}$ for all k. Hence

$$\lim_{k\to\infty} \sigma_k = \bar{\sigma} \geq 0.$$

Assume that $\bar{\sigma} > 0$. Then there exists a $\hat{k}$ such that

$$\bar{\sigma} \leq \sigma_{\hat{k}} < \frac{\bar{\sigma}}{\beta}$$

since $0 < \beta < 1$. Hence for $k \geq j \geq \hat{k}$,

$$\bar{\sigma} \leq \sigma_k \leq \sigma_j < \frac{\bar{\sigma}}{\beta}\,.$$

Thus $\beta\sigma_j < \sigma_k$ for all $k \geq j \geq \hat{k}$.

Consequently condition (b) of Deletion Rule 2 is never satisfied for $j \geq \hat{k}$. Two cases must be considered.

Case 1: $\min_{t\in S}[u^T(t)x_j - \lambda(t)] < 0.$

In this case Step 3(ii) is used to generate a cut and so for all $k \not> j \geq \hat{k}$

$$u^T(t_j)x_k - \|u(t_j)\|\sigma_k \geq \lambda(t_j). \tag{3}$$

From (2), we have

$$u^T(t_j)x_j \leq \lambda(t_j). \tag{4}$$

Subtracting (4) from (3) yields

$$u^T(t_j)\ (x_k - x_j) - \|u(t_j)\|\sigma_k \geq 0,$$

but this implies that

$$\|u(t_j)\|(\|x_k - x_j\| - \sigma_k) \geq 0. \tag{5}$$

Now $\|u(t_j)\| \neq 0$, since otherwise the contradiction

$$-\lambda(t_j) = u^T(t_j)x_j - \lambda(t_j) \leq 0,$$

and, using $\hat{x}$ from (vi),

$$-\lambda(t_j) = u^T(t_j)\hat{x} - \lambda(t_j) > 0$$

would follow. Consequently (5) yields

$$\|x_k - x_j\| \geq \sigma_k \geq \bar{\sigma} \quad \text{for all} \quad k > j \geq \hat{k}.$$

Case 2: $u^T(t)x_j - \lambda(t) \geq 0 \quad \forall t \in S.$
In this case

$$c^T x_k + \|c\|\sigma_k \leq c^T x_j \quad \text{for all} \quad k > j.$$

Hence

$$\|c\| \cdot \|x_k - x_j\| \geq \|c\|\sigma_k \quad \text{for all} \quad k > j.$$

Therefore in either case

$$\|x_k - x_j\| \geq \sigma_k \geq \bar{\sigma} \quad \text{for all} \quad k > j \geq \hat{k}.$$

This contradicts the fact that $\{x_k\}_{k=1}^{\infty} \subset H$ must have limit points since H is compact. Thus $\bar{\sigma} = 0$. $\|$

Lemma 2: If $\underset{\sim}{x}$ is feasible for Program D and $u^T(t)\underset{\sim}{x} > \lambda(t)$ for all $t \in S$, there exists a $\underset{\sim}{\sigma} > 0$ such that $(\underset{\sim}{x}, \underset{\sim}{\sigma})$ satisfies any cut generated by Step 3(ii).

Proof: Since $u(\cdot)$ and $\lambda(\cdot)$ are continuous on the compact set S and $u^T(t)\underset{\sim}{x} - \lambda(t) > 0$ for all $t \in S$, it follows that

$$\min_{t \in S} [u^T(t)\underset{\sim}{x} - \lambda(t)] = \epsilon > 0.$$

Consequently

$$u^T(t_k)\underset{\sim}{x} - \lambda(t_k) - \|u(t_k)\|\sigma \geq \epsilon - \|u(t_k)\|\sigma \quad \text{for all } k.$$

By assumptions (ii), (iv), and (vi) of Section 1, $\max_{t \in S} \|u(t)\| = M$ with $0 < M < \infty$. Choosing $\underset{\sim}{\sigma} \in (0, \frac{\epsilon}{M})$ we have that

$$u^T(t_k)\underset{\sim}{x} - \|u(t_k)\|\underset{\sim}{\sigma} \geq \lambda(t_k) \quad \text{for all } k. \quad \|$$

Lemma 3. If the algorithm stops at iteration k^*, y_{k^*-1} is feasible for Program D. If the algorithm does not terminate, there exists $\hat{k}$ such that y_k is feasible for Program D for $k \geq \hat{k}$.

Proof: Assume that for all k, y_k is not feasible for Program D. Thus for each k, x_k is not feasible for Program D. Hence Step 3(i) is never used to generate a constraint and Deletion Rule 1 is never used to drop a constraint.

Let x^* be optimal for Program D and let $\hat{x}$ be the point in assumption (vi). Then $x(\alpha) = \alpha\hat{x} + (1-\alpha)x^*$ is feasible for $0 \leq \alpha \leq 1$ and

$$u^T(t)x(\alpha) > \lambda(t) \qquad \forall t \in S \quad \text{for} \quad 0 < \alpha \leq 1. \tag{6}$$

Since $c^T x^* < \bar{b}$, we can choose $\underset{\sim}{\alpha} \in (0,1)$ sufficiently small such that $c^T x(\underset{\sim}{\alpha}) < \bar{b}$. Hence there exists $\underset{\sim}{\sigma}_1 > 0$ such that

$$c^T x(\alpha) + \|c\|\underset{\sim}{\sigma}_1 \leq \bar{b}.$$

By (6) and Lemma 2, there exists $\underset{\sim}{\sigma}_2 > 0$ such that $(x(\underset{\sim}{\alpha}), \underset{\sim}{\sigma}_2)$ satisfies any cut generated by Step 3(ii). Choosing $\underset{\sim}{\sigma} = \min(\underset{\sim}{\sigma}_1, \underset{\sim}{\sigma}_2) > 0$, $(x(\underset{\sim}{\alpha}), \underset{\sim}{\sigma})$ is feasible for SD_k for all k. Thus $\lim_{k \to \infty} \sigma_k \geq \underset{\sim}{\sigma} > 0$, a contradiction of Lemma 1. $\|$

Theorem 1: If the algorithm terminates at iteration k^*, then y_{k^*-1} is optimal for Program D. Otherwise, limit points exist to the sequence $\{y_k\}_{k=0}^{\infty}$ and they are optimal.

Proof: Suppose the algorithm does not terminate. By Lemmas 1 and 3 there exists a

$\hat{k}$ such that y_k is feasible for Program D for all $k \geq \hat{k}$. Since the feasible region for Program D is compact, limit points exist to $\{y_k\}_{k=0}^{\infty}$ and they are feasible. Let $\bar{y}$ be a limit point and $\{y_k\}_{k\in T}$ be a subsequence converging to $\bar{y}$. Suppose that $\bar{y}$ is not optimal. Let x^* be optimal and $\hat{x}$ the point specified in (vi). Then defining $x(\alpha) = \alpha x^* + (1-\alpha)\hat{x}$,

$$u^T(t)x(\alpha) > \lambda(t) \qquad \forall t\in S \quad \text{for} \quad 0 \leq \alpha < 1$$

and

$$c^T x(\alpha) < c\bar{y} \quad \text{for} \quad \max \left\{\frac{c^T\bar{y}-c^T\hat{x}}{c^T x^*-c^T\hat{x}}, 0\right\} < \alpha < 1.$$

Choose $\underset{\sim}{\alpha}\in\left(\max\left\{\frac{c^T\bar{y}-c^T\hat{x}}{c^T x^*-c^T\hat{x}}, 0\right\}, 1\right)$ and set $\underset{\sim}{\sigma}_1 = \|c\|^{-1}(c^T\bar{y}-c^T x(\underset{\sim}{\alpha})) > 0$. Then $(x(\underset{\sim}{\alpha}), \underset{\sim}{\sigma}_1)$ satisfies $c^T x + \|c\|\sigma \leq c^T\bar{y} \leq c^T y_k$ for $k \geq \hat{k}$. By Lemma 2, there exists a $\underset{\sim}{\sigma}_2 > 0$ such that $(x(\underset{\sim}{\alpha}), \underset{\sim}{\sigma}_2)$ satisfies any cut generated by Step 3(ii). Thus $(x(\underset{\sim}{\alpha}), \underset{\sim}{\sigma})$ where $\underset{\sim}{\sigma} = \min(\underset{\sim}{\sigma}_1, \underset{\sim}{\sigma}_2) > 0$ is feasible for SD_k for all k. Hence $\lim_{k\to\infty} \sigma_k \geq \underset{\sim}{\sigma} > 0$ which contradicts Lemma 1.

If the algorithm terminates at iteration k^*, y_{k^*-1} is feasible for Program D by Lemma 3. If y_{k^*-1} is not optimal, an argument similar to the above shows that $\sigma_{k^*} > 0$. Hence the algorithm could not have terminated. ||

3. Convergence Rate

Elzinga and Moore were able to show a linear convergence rate for their algorithm. In a similar manner, we can show that the algorithm of Section 2 has a linear convergence rate. For the remainder of this paper, we will assume that H is a polyhedral set defined by a set of linear inequalities,

$$a_i^T x \geq b_i \qquad i = 1,2,\ldots,m.$$

Let us also assume that by stage k of the algorithm, a point feasible for Program D has been found and Step 3(i) has been used to generate a cut. (If this is not the case, $c^T y_{k-1}$ is replaced by $\bar{b}$ in the following developments.) Under these assumptions, Program SD_{k-1} can be written as the following linear program.

$$\begin{aligned} &\underline{\max}\ \sigma \\ &\underline{\text{subject to}} \\ &c^T x + \|c\|\sigma \leq c^T y_{k-1} \\ &u^T(t_j)x - \|u(t_j)\|\sigma \geq \lambda(t_j) \qquad j\in\Omega_k \\ &a_i^T x \geq b_i \qquad i = 1,\ldots,m. \end{aligned}$$

In this program, Ω_k is the set of indices of the constraints that have been generated by Step 3(ii) but have not been dropped by Deletion Rule 2 before the k-th iteration. Also, we need only consider the last constraint generated by Step 3(i).

The dual program to Program SD_{k-1} is

<u>Program SP_{k-1}</u>:

$$\underline{\min}\ c^T y_{k-1} w_o - \sum_{j\in\Omega_k} \lambda(t_j) w_j - \sum_{i=1}^{m} b_i v_i$$

<u>subject to</u>

$$cw_o - \sum_{j\in\Omega_k} u(t_j) w_j - \sum_{i=1}^{m} a_i v_i = 0 \qquad (7)$$

$$\|c\| w_o + \sum_{j\in\Omega_k} \|u(t_j)\| w_j = 1$$

$$w_j \geq 0 \quad \text{for} \quad j\in\{0\} \cup \Omega_k$$

$$v_i \geq 0 \quad \text{for} \quad i = 1,2,\ldots,m. \quad \|$$

In the following developments, let (x_k, σ_k) be optimal for Program SD_{k-1} and let

$$w_j = w_j^k \quad \text{for} \quad j\in\{0\} \cup \Omega_k$$

$$v_i = v_i^k \quad \text{for} \quad i = 1,\ldots,m$$

be optimal for Program SP_{k-1}.

<u>Lemma 4</u>: Let $x\in\mathbb{R}^n$ satisfy the constraints of Program D. Then if $w_o^k > 0$,

$$c^T x \geq c^T y_{k-1} - \frac{\sigma_k}{w_o^k}.$$

<u>Proof</u>: Using the duality of Program SD_{k-1} and Program SP_{k-1} together with (7) we have

$$\begin{aligned}\sigma_k &= c^T y_{k-1} w_o^k - \sum_{j\in\Omega_k} \lambda(t_j) w_j^k - \sum_{i=1}^{m} b_i v_i^k \\ &\quad - x^T (c w_o^k - \sum_{j\in\Omega_k} u(t_j) w_j^k - \sum_{i=1}^{m} a_i v_i^k) \\ &= - w_o^k (c^T x - c^T y_{k-1}) + \sum_{j\in\Omega_k} w_j^k (u^T(t_j) x - \lambda(t_j)) \\ &\quad + \sum_{i=1}^{m} v_i^k (a_i^T x - b_i) \end{aligned} \qquad (8)$$

Since x is feasible for Program D,

$$u^T(t)x \geq \lambda(t) \quad \text{for all} \quad t \in S \tag{9}$$

$$\text{and} \quad a_i^T x \geq b_i \qquad i = 1,2,\ldots,m. \tag{10}$$

Combining (8), (9), and (10) and the fact that

$$w_j^k \geq 0 \text{ for all } j \text{ and } v_i^k \geq 0 \text{ for } i = 1,\ldots,m,$$

we have that

$$\sigma_k \geq - w_o^k(c^T x - c^T y_{k-1}). \quad \|$$

As in [3], this gives the bound

$$c^T y_{k-1} - \frac{\sigma_k}{w_o^k} \leq V_D \leq c^T y_k \tag{11}$$

when $w_o^k > 0$.

Lemma 5: Assume that the algorithm does not terminate and let $\underline{w}_o = \liminf_{k \to \infty} w_o^k$ and $\bar{w}_o = \limsup_{k \to \infty} w_o^k$. Then

$$0 < \underline{w}_o \leq \bar{w}_o < \infty .$$

If the algorithm terminates at stage k^*, then $w_o^{k^*} > 0$.

Proof: From Program SP_{k-1} we have

$$\|c\| w_o^k + \sum_{j \in \Omega_k} \|u(t_j)\| w_j^k = 1 \tag{12}$$

$$w_o^k \geq 0; \; w_j^k \geq 0 \qquad j \in \Omega_k. \tag{13}$$

Thus $0 \leq w_o^k \leq \frac{1}{\|c\|}$ for all k, and so $0 \leq \underline{w}_o \leq \bar{w}_o < \infty$. We must now show that $\underline{w}_o$ is strictly positive.

From assumptions (ii), (iv) and (vi) we have that

$$\max_{t \in S} \|u(t)\| = M$$

where $0 < M < \infty$. Combining this with (12) and (13) yields

$$\sum_{j\in\Omega_k} w_j^k \geq \frac{1 - \|c\| w_o^k}{M} \tag{14}$$

Let $\hat{x}$ be the point in assumption (vi); then

$$\min_{t\in S}(u^T(t)\hat{x} - \lambda(t)) = q > 0.$$

Using this relation and (8) yields

$$\sigma_k \geq - w_o^k(c^T\hat{x} - c^T y_{k-1}) + q \sum_{j\in\Omega_k} w_j^k.$$

Combining this with (14), we have

$$\sigma_k \geq - w_o^k(c^T\hat{x} - c^T y_{k-1}) + \frac{q}{M}(1 - \|c\| w_o^k).$$

Now $V_D \leq c^T y_{k-1}$, so

$$w_o^k(c^T\hat{x} - V_D + \frac{q\|c\|}{M}) \geq - \sigma_k + \frac{q}{M}.$$

By assumption (vi), $c^T\hat{x} - V_D > 0$; thus

$$c^T\hat{x} - V_D + \frac{q\|c\|}{M} \equiv h > 0.$$

Hence

$$w_o^k \geq - \frac{\sigma_k}{h} + \frac{q}{Mh}$$

Using Lemma 1 and taking limits,

$$\underline{w}_o \geq \frac{q}{Mh} > 0.$$

If the algorithm terminates at k^*, $\sigma_{k^*} = 0$ and so

$$w_o^{k^*} \geq \frac{q}{Mh} > 0. \quad \|$$

Theorem 2: Between feasible points, the algorithm has a linear rate of convergence in objective function value.

Proof: From Lemma 5, we have that $\underline{w}_o > 0$. This and Lemma 3 shows that there exists a $\hat{k}$ such that for $k \geq \hat{k}$, w_o^k is positive and y_{k-1} is feasible for Program D. Since Programs SD_{k-1} and SP_{k-1} are dual, this result and complementary slackness imply

$$c^T x_k + \|c\|\sigma_k = c^T y_{k-1}.$$

Using Lemma 4 and the above result we have for $k \geq \hat{k}$,

$$V_D \geq c^T y_{k-1} - \frac{c^T y_{k-1} - c^T x_k}{\|c\| w_o^k}.$$

Thus there is a $\rho \in (0,1)$ such that for k sufficiently large

$$\begin{aligned} c^T y_{k-1} - c^T x_k &\geq \|c\| w_o^k (c^T y_{k-1} - V_D) \\ &\geq \rho \|c\| \underline{w}_o (c^T y_{k-1} - V_D). \end{aligned} \tag{15}$$

If x_k is feasible for Program D, $y_k \equiv x_k$. Thus for k sufficiently large, (15) yields

$$\frac{c^T y_k - V_D}{c^T y_{k-1} - V_D} \leq 1 - \rho \|c\| \underline{w}_o$$

when $y_k \neq y_{k-1}$. ||

4. The Dual Program

The following program is dual to Program D with H given as in Section 3.

Program P:

Find $V_P = \sup \sum_{t \in S} \lambda(t)\xi(t) + \sum_{i=1}^{m} b_i \nu_i$

subject to

$$\sum_{t \in S} u(t)\xi(t) + \sum_{i=1}^{m} a_i \nu_i = c$$

$\nu_i \geq 0$ for $i = 1,2,\ldots,m$

$\xi(t) \geq 0$ for all $t \in S$ and

$\xi(t) = 0$ for all but a finite number of t in S. ||

It is easy to show that

$$V_P \leq V_D. \tag{16}$$

Theorem 3: Let

$$w_j = w_j^k \quad \text{for} \quad j \in \{0\} \cup \Omega_k$$
$$v_i = v_i^k \quad \text{for} \quad i = 1,\ldots,m$$

be optimal for Program SP_{k-1}. If $w_o^k \neq 0$, define

$$\xi^k(t) = \begin{cases} w_j^k/w_o^k & \text{for } t \in \{t_j \mid j \in \Omega_k\} \\ 0 & \text{otherwise} \end{cases}$$

and

$$\nu_i^k = v_i^k/w_o^k \quad \text{for} \quad i = 1,2,\ldots,m.$$

Then

$$\xi(t) = \xi^k(t)$$

$$\nu_i = \nu_i^k \qquad i = 1,\ldots,m$$

is feasible for Program P. Also,

$$\lim_{k\to\infty} \sum_{t \in S} \lambda(t)\xi^k(t) + \sum_{i=1}^{m} b_i\nu_i^k = V_P$$

if the algorithm does not terminate.

If the algorithm terminates at stage k^*, then

$$\xi^{k^*}(t);\ \nu_i^{k^*} \qquad i = 1,\ldots,m$$

is optimal for Program P.

Proof: Assume that the algorithm does not terminate. From Program SP_{k-1}, we have

$$w_j^k \geq 0 \quad \text{for} \quad j \in \{0\} \cup \Omega_k$$

$$v_i^k \geq 0 \quad \text{for} \quad i = 1,2,\ldots,m.$$

Thus if $w_o^k \neq 0$,

$$\xi^k(t) \geq 0 \quad \text{for all} \quad t \in S \quad \text{and}$$
$$\xi^k(t) = 0 \quad \text{unless} \quad t \in \{t_j \mid j \in \Omega_k\} \tag{17}$$
$$\nu_i^k \geq 0 \quad \text{for} \quad i = 1,2,\ldots,m.$$

From (7) we have

$$cw_o^k - \sum_{j\in\Omega_k} u(t_j)w_j^k - \sum_{i=1}^m a_i v_i^k = 0$$

Dividing through by $w_o^k \neq 0$,

$$c - \sum_{t\in S} u(t)\xi^k(t) - \sum_{i=1}^m a_i \nu_i^k = 0 \tag{18}$$

Relations (17) and (18) show that

$$\xi(t) = \xi^k(t) \quad \text{for} \quad t\in S$$
$$\nu_i = \nu_i^k \qquad i = 1,2,\ldots,m$$

is feasible for Program P.

From Lemma 5, we have that there is a $\hat{k}$ such that for $k \geq \hat{k}$

$$w_o^k > 0.$$

Thus for $k \geq \hat{k}$, we can find $\xi^k(\cdot)$ and ν_i^k. Also, by Lemma 3, we can assume that $\hat{k}$ is large enough so that y_{k-1} is feasible for Program D if $k \geq \hat{k}$.

Using the duality between Programs SP_{k-1} and SD_{k-1}, we have for $k \geq \hat{k}$,

$$\sigma_k = c^T y_{k-1} w_o^k - \sum_{j\in\Omega_k} \lambda(t_j)w_j^k - \sum_{i=1}^m b_i v_i^k .$$

Since $k \geq \hat{k}$, we can divide through by w_o^k,

$$\sum_{t\in S} \lambda(t)\xi^k(t) + \sum_{i=1}^m b_i \nu_i^k = c^T y_{k-1} - \frac{\sigma_k}{w_o^k} . \tag{19}$$

From Lemma 1, Lemma 5, Theorem 1, (16) and (19), we have

$$V_P \geq \lim_{k\to\infty} \sum_{t\in S} \lambda(t)\xi^k(t) + \sum_{i=1}^m b_i \nu_i^k = V_D \geq V_P.$$

If the algorithm terminates, we have $w_o^{k*} > 0$ by Lemma 5. Thus (17) and (18) still hold, showing feasibility. Using (19), $\sigma_{k*} = 0$, and Theorem 1, we have

$$V_P \geq \sum_{t\in S} \lambda(t)\xi^{k*}(t) + \sum_{i=1}^m b_i \nu_i^{k*} = c^T y_{k*-1} = V_D \geq V_P. \quad \|$$

5. Concluding Remarks

The algorithm given in this paper has the property that, if at any iteration a point that is infeasible for the semi-infinite program is generated, a cut can be developed from any violated or tight constraint. Theorem 1 shows that the algorithm generates a sequence of points whose limit points are optimal, and Theorem 2 shows that there is a linear rate of convergence. However, this leaves open the question of whether generating the cut from the most violated constraint would significantly improve the computational performance of the algorithm.

In [7], we studied the use of central-cutting-plane algorithms on primal prototype geometric programs. The Elzinga and Moore algorithm is applied to the convex program obtained from the geometric program by logarithmic transformation of variables. The algorithm of this paper is applied to an equivalent semi-infinite program [1]. If an infeasible point is given and the most violated constraint of the semi-infinite program is used to generate a cut, the resulting cut is strictly deeper than the one generated by the Elzinga and Moore algorithm from the most violated constraint of the convex program. Using the respective most violated constraint in each algorithm, the Elzinga and Moore algorithm required approximately three percent more iterations to obtain a point within a given percent of optimal than needed by the semi-infinite algorithm.

In [5], Gochet and Smeers apply the alternating algorithm to the semi-infinite program and the Kelley algorithm to the convex program. They show that the alternating algorithm generates strictly deeper cuts from a given infeasible point. In this case, the Kelley algorithm requires approximately three <u>times</u> as many iterations as the alternating algorithm to obtain a given degree of feasibility, see [2]. This tends to indicate that the semi-infinite central-cutting-plane algorithm is relatively insensitive to the depth of the cuts generated. Thus it may not be necessary to find the most violated constraint to obtain good performance.

In [13] and [14], Gustafson and Kortanek use duality relations between Programs P and D to develop a system of nonlinear equations in the primal and dual variables that are necessary and sufficient conditions for primal and dual feasible solutions to be optimal solutions. They also show that we need only consider solutions to Program P with at most n variables taking non-zero value.

The system of equations is of the form:

System NL:

$$\sum_{i=1}^{n} u(t_i)\xi(t_i) + \sum_{i=1}^{m} a_i \nu_i = c$$

$$\xi(t_i)\,(u^T(t_i)x - \lambda(t_i)) = 0 \qquad i = 1,\ldots,n$$

$$\nu_i(a_i^T x - b) = 0 \qquad i = 1,\ldots,m$$

$(u^T(t)x-\lambda(t))$ has a local minimum at t_i if $\xi(t_i) > 0$.
The variables in this system are

$$\xi(t_i),\ i = 1,\dots,n;\ t_i \in S,\ i = 1,\dots,n;\ x \in \mathbb{R}^n;$$
$$\nu_i,\ i = 1,\dots,m \quad (\xi(t) = 0 \text{ if } t \neq t_i).$$

They propose to find a trial solution as input to System NL and to use Newton-Raphson iterations. If the Newton-Raphson procedure converges, they check to see if the results are feasible for Programs P and D in which case they are also optimal. Since we are using Newton-Raphson, the convergence may be quadratic under certain conditions; however, convergence is not guaranteed.

They find their trial solutions by replacing S with a grid T containing a finite number of points and solving the resulting linear programs. The solutions of these linear programs are used as the input for the Newton-Raphson iterations in System NL. If the Newton-Raphson procedure converges to feasible points, optimal solutions have been found; if not, a finer grid is used and the process is repeated. The problem with this procedure is that very large linear programs may have to be solved before the Newton-Raphson procedure converges to feasible solutions.

Our algorithm produces a sequence of feasible points for Program D whose limit point are optimal points. Also, in Theorem 3, we showed that we can produce a sequence of feasible solutions to Program P whose values converge to V_P. (It is easy to show that we can "refine" these solutions so that at most n variables have non-zero values.) Thus we can use our algorithm to produce initial points for Newton-Raphson iterations on System NL. The algorithm of Section 2 can be run until feasible solutions for Programs P and D are found. Newton-Raphson procedure starting with these points can then be used to try to solve System NL. If the procedure converges to feasible points, we are done; otherwise, continue with the cutting-plane algorithm. Such a procedure could be well suited to parallel processing. One processor would run the cutting-plane algorithm. The other would use the feasible points found to run Newton-Raphson iterations on System NL and check for convergence.

REFERENCES

[1] A. Charnes, W. W. Cooper and K. O. Kortanek, "Semi-Infinite Programming Differentiability and Geometric Programming. Part I: With Examples and Applications in Economics and Management Science," R. S. Varma Memorial Volume, J. Math. Sciences of India 6 (1971), pp. 19-40.

[2] John J. Dinkel, William H. Elliott, Gary A. Kochenberger, "Computational Aspects of Cutting-Plane Algorithms for Geometric Programming Problems," Mathematical Programming 13 (1977), pp. 200-220.

[3] J. Elzinga and T. G. Moore, "A Central Cutting Plane Algorithm for the Convex Programming Problem," Mathematical Programming, Vol. 8 (1975), pp. 134-145.

[4] K. Fahlander, "Computer Programs for Semi-Infinite Optimization," TRITA-NA-7312, Department of Numerical Analysis, Roval Institute of Technology, S-10044, Stockholm 70, Sweden.

[5] W. Gochet, and Y. Smeers, "On the Use of Linear Programs to Solve Prototype Geometric Programs," CORE Discussion Paper No. 7229, 1972.

[6] P. R. Gribik and K. O. Kortanek, "Equivalence Theorems and Cutting Plane Algorithms for a Class of Experimental Design Problems," SIAM J. Appl. Math., Vol. 32, (1977), pp. 232-259.

[7] P. R. Gribik and D. N. Lee, "A Comparison of Central-Cutting-Plane Algorithms for Prototype Geometric Programming Problems," presented at III Symposium über Operations Research, Mannheim (1978).

[8] P. R. Gribik, "Selected Applications of Semi-Infinite Programming," Department of Mathematics, Carnegie-Mellon University, presented at "Constructive Approaches to Mathematical Models": A Symposium in Honor of R. J. Duffin, Pittsburgh (1978).

[9] S.-Å. Gustafson, "On the Computational Solution of a Class of Generalized Moment Problems," SIAM. J. Numer. Anal., Vol. 7 (1970), pp. 343-357.

[10] S.-Å. Gustafson, "Nonlinear Systems in Semi-Infinite Programming," Solutions of Nonlinear Algebraic Equations, edited by G. B. Byrnes and C. A. Hall, Academic Press, 1973.

[11] S.-Å. Gustafson, "On Computational Applications of the Theory of the Moment Problem," The Rocky Mt. J. of Math., Vol. 4 (1974), pp. 227-240.

[12] S.-Å. Gustafson and K. O. Kortanek, "Numerical Solution of a Class of Convex Programs," Meth. of Op. Res., Vol. XVI (1972), pp. 138-149.

[13] S.-Å. Gustafson and K. O. Kortanek, "Numerical Solution of a Class of Semi-Infinite Programming Problems," NRLQ, Vol. 20 (1973), pp. 477-504.

[14] S.-Å. Gustafson and K. O. Kortanek, "Computational Schemes for Semi-Infinite Programs," Technical Report, Department of Mathematics, Carnegie-Mellon University, Pittsburgh, PA.

[15] J. E. Kelley, "The Cutting Plane Method for Solving Convex Programs," J. SIAM, Vol. 8 (1960), pp. 703-712.

A STABLE MULTIPLE EXCHANGE ALGORITHM FOR LINEAR SIP

By Klaus Roleff

The subject of linear optimization belongs to those branches in mathematics which are widely applied in economics and natural sciences. Especially a "good" solving of linear optimization problems on a computer is of importance today and users are looking for fast and stable methods. So it gives much reason for further research in this subject.

There are different methods to compute an optimal solution of a linear optimization problem, of which the exchange algorithms (simplex method by DANTZIG [51]) and the descent algorithms (gradient methods) are the most important ones. And it seems today that the exchange methods have outrun the descent methods cosidering computing times and exactness of the solutions.

The fundamental exchange algorithm for solving linear SIP is the simplex method. And the multiple exchange algorithm is a generalization of the simplex procedure, allowing more than one vector per iteration step to enter the basis.

The crucial point for the execution of exchange algorithms for linear SIP is the solving of linear equations. Solution methods should be numerical stable: A use of a very time and storage sparing method has at least no sense, if this method can even in stable problems deliver solutions, which are of no use because of heavy rounding errors.

We note that all other exchange methods for linear SIP, considering their main idea, can be subordinated to the simplex algorithm. It can even be shown that with few exceptions these methods deliver the same iteration solutions as the simplex method or the multiple exchange procedure. By specialisation on special problem classes the execution of the simplex or multiple exchange iterations may be simplified.

So it seems convenient to give a brief summary of the simplex procedure, whereby we will follow essentially the recently published book by GLASHOFF/GUSTAFSON [78]. The multiple exchange algorithm later will be treated total analogously.

§ 1 The Simplex Method

The linear problems to be solved may have the following standard form. Be given a set $S \subset R^l$, $l \geq 1$, S not empty and not necessary finite, real functions $a_r(s)$, $r=1,..,n$ and $b(s)$, $s \in S$, and a fixed vector $c = (c_1,...,c_n)^T$.
The search is for vectors $y = (y_1,...,y_n)^T$ and $x = (x_1,...,x_n)^T$ to solve

$$\text{Minimize} \quad \sum_{r=1}^{n} y_r c_r \qquad\qquad \text{(P)}$$

under the restrictions (primal problem)

$$(1) \qquad \sum_{r=1}^{n} y_r a_r(s) \geq b(s) \quad \forall s \in S$$

$$\text{Maximize} \quad \sum_{i=1}^{n} x_i b(s_i) \qquad\qquad \text{(D)}$$

under the restrictions (dual problem)

$$(2) \qquad \sum_{i=1}^{n} x_i a_r(s_i) = c_r \qquad r=1,..,n$$

$$(3) \qquad x_i \geq 0 \quad i=1,..,n,$$

$$s_i \in S \quad i=1,..,n$$

For all to come we make the following <u>assumptions</u> :

I) (P) and (D) have no duality gap - that is the optimal values of (P) and (D) are equal

II) (D) is solvable or unbounded

III) There are n linearly independent vectors in $\{a(s),\ s \in S\}$ with $a(s) = (a_1(s),...,a_n(s))^T$.

<u>definition</u>: Be $\mathcal{G} := \{s_1,...,s_n\} \subset S$. We call $\{\mathcal{G}, x\}$ <u>basis solution</u>, if the vectors $a(s_1),...,a(s_n)$ are linearly independent and x is consistent with (D) - that is x solves (2) and (3). Thus x is defined by the linear system

$$(4) \qquad A(s_1,...,s_n)x = c$$

where $A(s_1,...,s_n) := (a(s_1),...,a(s_n))$ is the <u>basis matrix</u>. $\mathcal{G}$ is designated as <u>basis set</u> and the vectors $a(s_i)$, $i=1,..,n$ we mark as <u>basis</u>.

We remark that because of the generell assumptions (D) has an optimal basis solution if solvable. Therefore in the formulation of (D) we have summed up only to n.

If $\mathfrak{S}$ is a basis set the corresponding approximate solution for the primal problem is defined in the simplex method by the balance theorem - that means by solving

(5) $\quad A(s_1,\ldots,s_n)^T y \;=\; (b(s_1),\ldots,b(s_n))^T.$

Starting with the basis solution $\{x,\mathfrak{S}\}$ the aim in each simplex step is to get a new basis solution $\{\tilde{x},\tilde{\mathfrak{S}}\}$ corresponding to a greater value of the target function. The transition from one basis solution to another can be most easily described, when only one basis vector per iteration step is exchanged. So is the main question how the solution of (4) changes, when one column of $A(s_1,\ldots,s_n)$ is exchanged.

This process is known:
Assume that an element $s^* \in S$ shall enter the basis set, then first solve the set of linear equations (6) to get a representation of the vector $a(s^*)$:

(6) $\quad A(s_1,\ldots,s_n)d \;=\; a(s^*) \qquad\qquad (d = (d_1,\ldots,d_n)^T).$

When we set

(7) $\quad x(\lambda) = (x_1-\lambda d_1,\ldots,x_n-\lambda d_n,\lambda)^T$

we know that $x(\lambda)$ solves the equality restrictions (2) of (D) and to ensure for all components $x_i(\lambda) \geq 0$, $i=1,\ldots,n$, we chose λ as

(8) $\quad \bar{\lambda} = \frac{x_r}{d_r} = \min_{i=1,\ldots,n} \left\{\frac{x_i}{d_i} \;\middle|\; d_i > 0\right\} \qquad \text{with } r \in \{1,\ldots,n\}.$

If there is no $d_i > 0$ then (D) is unbounded and (P) not consistent.

If the minimum in (8) is attained for an index r, the element s_r is removed from the basis set. The new objective function value for $x(\bar{\lambda})$ will be

(9) $\quad c_o(\bar{\lambda}) \;=\; \underbrace{c_o(0)}_{=\text{value for } x} \;+\; \bar{\lambda}(b(s^*) - \sum_{r=1}^{n} y_r a_r(s^*))$

and this shows that achieving an increase of this value, s^* must be chosen so that the primal restrictions (1) are violated in s^*. If such an s^* does not exist, x and y are optimal.

Therefore we have two main tasks in the simplex method,

1) to solve three sets of linear equations (4),(5) and (6).
2) to test the primal restrictions in order to get a "good" s^*.

In § 2 we want to make some remarks on the more critical part thereof - the first task.

§ 2 Stable Matrix Decompositions And Modification Methods

It is obvious, especially for greater n, that it is a time spending task to solve in each simplex step three systems of linear equations. So it is desired to use quick but stable methods.

Since the matrix of all three sets of linear equations is $A(s_1,\dots,s_n)$ or $A(s_1,\dots,s_n)^T$, matrix decompositions can be used to solve these linear systems with less work than a repeatedly application of - say the Gauß elimination method.

Performing one simplex step the matrix defining the linear systems (4), (5) and (6) is changed in only one column or row. The so called modification methods make use of this fact. Solving a linear system with one of the elimination methods, Householder or Givens partitions, the work is proportional to n^3 operations (multiplications and additions) and this is reduced with modification methods to a work proportional only to n^2 operations.

For the simplex method one of the following methods is usually used:

a) Gauß Jordan elimination
b) explicite use of the inverse basis matrix

The corresponding modification methods - in this case the so called recursion formulas - are not stable.

Stable decomposition methods with corresponding stable modification methods are

c) Gauß elimination with column pivot search
d) Householder partitions
e) Givens partitions

Literature references : BARTELS/GOLUB [68]and [69] GILL/MURRAY [73], GILL/GOLUB/MURRAY/SAUNDERS [74], DANIEL/GRAGG/KAUFMANN/STEWART [76], STOER [76] and GLASHOFF/GUSTAFSON [78].

Using the Gauß algorithm for computing a matrix decomposition is less time consuming than the two other stable methods and it can be most easily carried out. It works the following way (see GLASHOFF/GUSTAFSON [78]):

Applying the ordinary Gauß elimination process with column pivot search to the $n \times 2n$ matrix $(A|I)$, I = identity matrix, you get $(R|F)$ where R is an upper $n \times n$ trianqular matrix and F an $n \times n$ matrix so that $FA = R$. On the other side, having R and F, the sets of linear equations (4), (5) and (6) can be easily solved:

for example (4) $\quad Ax = c \iff FAx = Fc$

(10) $\quad \iff Rx = Fc$

and (10) is solved by "backward substitution". (5) and (6) can be solved quite similiar. The work for solving a linear system in this manner is proportional to n^2 operations.

Of course you don't want to make in each simplex step a wholly new partition of the basis matrix (number of operations prop. n^3), this may be more economically done with the corresponding modification method:
If $A = (a(s_1),\ldots,a(s_n))$ and $\tilde{A} = (a(s_1),\ldots,a(s_{r-1}),a(s_{r+1}),\ldots,a(s_n),a(s^*))$, form

$F\tilde{A}$ =

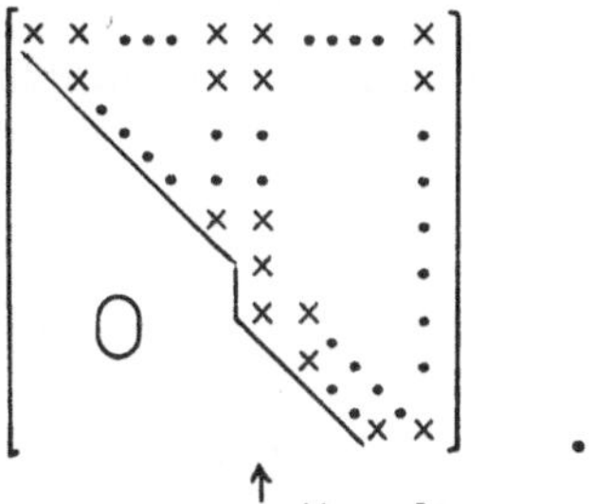

r-th column

Applying a simplified Gauß method to $(F\tilde{A}|F)$ (in searching the greatest column pivot element, you have to compare only two rows), you get the new partition for $\tilde{A}$: $(\tilde{R}|\tilde{F})$ and now the number of operations is proportional only to n^2.

§ 3 The Multiple Exchange Method

Some exchange algorithms solving linear Chebyshev approximation problems, allow more than one vector per iteration step to enter the basis, see for example the Remez multiple exchange algorithm. So there naturally is the question, whether it is possible to develope a simplex multiple exchange method. Such an algorithm for finite linear SIP problems indeed can be found in the book of JUDIN/GOLSTEIN [68].

In the description of this method let us follow the same concept as for the single exchange procedure. Let $\mathfrak{S} = \{s_1,\ldots,s_n\}$, then chose $k \geq 1$, $k \in \mathbb{N}$ and let the vectors $a(s_1^*),\ldots,a(s_k^*)$ be permitted to enter the basis. The exchange should be executed so that the result is again a basis solution.

First we need a representation of the vectors $a(s_i^*)$, $i=1,..,k$ by solving

(11) $\quad A(s_1,\ldots,s_n)d^i = a(s_i^*) \qquad i=1,\ldots,k \qquad (\text{with } d^i = (d_1^i,\ldots,d_n^i)^T)$.

It can easily be shown that

$$x(\lambda) := x(\lambda_1,\ldots,\lambda_k) = (x_1 - \sum_{i=1}^{k} \lambda_i d_1^i,\ldots\ldots,x_n - \sum_{i=1}^{k} \lambda_i d_n^i,\lambda_1,\ldots\ldots,\lambda_k)^T$$

fulfills the equality restrictions (2) of (D). The value $c_o(\lambda)$ belonging to $x(\lambda)$ can be calculated as

$$(13)\qquad c_o(\lambda) = c_o(0) + \sum_{i=1}^{k} \lambda_i \underbrace{\left(b(s_i^*) - \sum_{r=1}^{n} y_r a_r(s_i^*) \right)}_{:=\Delta(s_i^*)} .$$

If we require all $\Delta(s_i^*)$ to be greater zero, $i=1,..,k$, our aim is to exchange so, that the increase $c_o(\lambda)-c_o(0)$ of the objective function will be maximal and all the components of $x(\lambda)$ will be non negative. And this means we have to solve an finite auxiliary problem, which in the standard dual formulation has the following form:

$$\text{Maximize } \sum_{i=1}^{k} \lambda_i \Delta(s_i^*)$$

(14) under the restrictions

$$\begin{bmatrix} d_1^1 & \cdots & d_1^k \\ \vdots & & \vdots \\ d_n^1 & \cdots & d_n^k \end{bmatrix} \begin{bmatrix} \lambda_1 \\ \vdots \\ \lambda_k \end{bmatrix} + \begin{bmatrix} 1 & & 0 \\ & \ddots & \\ 0 & & 1 \end{bmatrix} \begin{bmatrix} \theta_1 \\ \vdots \\ \theta_n \end{bmatrix} = \begin{bmatrix} x_1 \\ \vdots \\ x_n \end{bmatrix}$$

$$\lambda_1 \geq 0, \ldots, \lambda_k \geq 0, \quad \theta_1 \geq 0, \ldots, \theta_n \geq 0.$$

Using the above described simplex method to solve (14) and assuming for simplicity that we have got an optimal basis solution of the form $(\lambda_1,\ldots,\lambda_{k'},\theta_{k'+1},\ldots,\theta_n)^T$, $k' \leq k$ (all other variables are equal to zero), the restrictions of (14) for this solution have the form

$$\begin{array}{lcl} d_1^1\lambda_1 + \cdots\cdots + d_1^{k'}\lambda_{k'} = x_1 & \Longleftrightarrow & \overbrace{x_1 - d_1^1\lambda_1 - \cdots - d_1^{k'}\lambda_{k'}}^{= x_1(\lambda)} = 0 \\ \vdots & & \vdots \\ d_{k'}^1\lambda_1 + \cdots\cdots + d_{k'}^{k'}\lambda_{k'} = x_{k'} & \Longleftrightarrow & \underbrace{x_{k'} - d_{k'}^1\lambda_1 - \cdots - d_{k'}^{k'}\lambda_{k'}}_{= x_{k'}(\lambda)} = 0 \\ d_j^1\lambda_1 + \cdots\cdots + d_j^{k'}\lambda_{k'} + \theta_j = x_j & \Longleftrightarrow & x_j(\lambda) = \theta_j \quad n \geq j > k' . \end{array}$$

We see that the first k' components of λ are in the basis, these are the components belonging to $a(s_1^*),\ldots,a(s_{k'}^*)$ and the first k' components of $x(\lambda)$ are vanishing, these are the components belonging to $a(s_1),\ldots,a(s_{k'})$. So we replace $s_1,\ldots,s_{k'}$ in $\mathcal{G}$ by $s_1^*,\ldots,s_{k'}^*$. One can show that on this manner a new basis solution for (D) is found.

Judin and Golstein work with the revised simplex method, which uses explicitly the inverse of the basis matrix to solve the linear systems (4) - (6). The new inverse they get in a difficult manner using the inverse basis matrix of the auxiliary problem (14).

To reduce the work for computing the inverse basis matrix during the iterations for solving (14), recursion formulas are used but these are numerically not stable.

§ 4 A Stable Multiple Exchange Algorithm

Working in the above described multiple exchange algorithm with stable matrix decompositions, there is the difficulty that the main problem (D) and the auxiliary problem (14) have totally different basis matrices, so that after solving the auxiliary problem no modification methods can be used to get the next decomposition of the usually in more than one vector changed basis matrix for (D). Therefore we shall reformulate (14) now.

Multiplicate the equality restrictions (14) by the actual basis matrix. Thereby the admissable region and therefore the problem itself is not changed. When we further consider that maximizing the increase of the objective function is equivalent of maximizing the objective function itself, the auxiliary problem now has the form:

$(\tilde{S} := \{s_1^*,\dots,s_k^*, s_1,\dots,s_n\}, \quad \hat{x}_1=\lambda_1,\dots,\hat{x}_k=\lambda_k, \ \hat{x}_{k+1}=\theta_1,\dots,\hat{x}_{k+n}=\theta_n)$

$$\text{Maximize} \sum_{i=1}^{k} \hat{x}_i b(s_i)$$

(15) under the restrictions

$$\sum_{i=1}^{n} \hat{x}_i a_r(s_i) = c_r \qquad r=1,\dots,n$$

$$\hat{x}_i \geq 0, \ i=1,\dots,n$$

$$s_i \in \tilde{S}, \ i=1,\dots,n$$

and this is nothing else than the original dual problem (D) restricted to $\tilde{S}$.

So we see: one step of the multiple exchange algorithm consists of some steps of the simplex method applied to a reduced finite problem. As starting solution for (15) we take x - the solution of the foregoing multiple step - and the decomposition of the corresponding basis matrix is known. Whenever a vector $a(s_i^*)$ enters the basis, we use stable modification methods to get quite simply a new partition of the changed basis matrix. Only after solving (15) we have to test all restrictions of (P) in order to find a new set $\tilde{S}$ defining the next auxiliary problem. So the execution of this algorithm is as simple as the simplex algorithm itself, the principle thereof is known in economics under the concept of multiple pricing, see for example ORCHARD-HAYS[68] .

We made a number of test computations to compare both algorithms,and a number of them are added at the end of this paper (§ 5). These computations seem to indicate that the multiple exchange algorithm normally needs less time for solving a linear SIP problem - a good choice of the s_i^* presupposed.

We suggest that at least for approximation problems the number of single exchanges in the multiple exchange procedure and the simplex algorithm are approximately the same. The advantage of the multiple exchange algorithm lies in the reduced work in testing the primal restrictions. Let us suggest for example that for solving a linear optimization problem with $m \geq 2n$ restrictions the simplex method need nr iterations. Equally the multiple exchange algorithm may have to solve for the same task r-times an auxiliary problem of the form (15), whereto each time n iterations are performed. In both cases then rn single exchange steps are executed but it can easily be calculated that in the multiple exchange procedure we have $r(mn^2-mn-2n^3)$ operations less for testing the primal restrictions than in the simplex method, and this means for $n=10$, $m=300$, 25000r operations !

Similiar the testing of the primal restrictions in the infinite case is more consumptuous than the testing in (15).

We remark that it is less work to execute the multiple exchange algorithm in the form presented in this chapter than in the form of § 3.

Cycling can be prevented by using the lexicographic rule, see for example KLOSTERMAIR[74] . And the convergence proof for the simplex method by KLOSTERMAIR[74] may also be used for the multiple exchange algorithm in this chapter, since this algorithm is nothing else than a sequence of simplex steps. For k=1 the multiple exchange procedure is identical with the simplex method - if the exchange in the not regular case (some basis variables equal to zero) is performed equally in both cases.

§ 5 Numerical Results

The multiple exchange algorithm in all examples was so computed that all local minima of the error function $\sum_{r=1}^{n} y_r a_r(s)-b(s)$ were chosen for the elements s_i^*, $i=1,..,k$, presupposed that the corresponding primal restrictions were violated. In the simplex method s^* denotes the element from S with the corresponding most violated restriction. In the infinite examples all elements s_i^*, i=1,..,k or s^* were computed with the Newton method.

All computations were made in FORTRAN on the TELEFUNKEN TR 440 computer of the computer centre of Hamburg university.

EXAMPLE 1 (simultaneous approximation), see BREDENDIEK/COLLATZ[76].

You want to have an approximate solution of

(a) $\Delta\Delta u(t,z) = 1 + t^2 + z^2$ for $(t,z) \in S = \{ (t,z) \mid |t| < 1, |z| < 1 \}$
(b) $u = 0$ and $\Delta u = 0$ on ∂S.

It is proved that for solutions u,v for (a) the following estimation is valid:

$$\text{(c)} \max_{(t,z)\in S} |(u-v)(t,z)| \leq \overbrace{\max_{(t,z)\in\partial S} |(u-v)(t,z)|}^{=:y_{n+1}} + \frac{1}{2} \overbrace{\max_{(t,z)\in\partial S} |(\Delta u-\Delta v)(t,z)|}^{=:y_{n+2}}.$$

Set $v(y) = v_0 + \sum_{i=1}^{6} y_i v_i$, v_i real functions on S, i=0,..,6, so that v(y) solves (a):

$$v_0(t,z) = \frac{1}{48}(t^4+z^4) + \frac{1}{360}(t^6+z^6)$$

$$v_1(t,z) = 1 \qquad v_2(t,z) = \frac{1}{4}(t^2+z^2) \qquad v_3(t,z) = t^4 - 6t^2z^2 + z^4$$

$$v_4(t,z) = \frac{1}{20}(t^6 - 5t^4z^2 - 5t^2z^4 + z^6) \quad v_5(t,z) = t^8 - 28t^6z^2 + 70t^4z^4 - 28t^2z^6 + z^8$$

$$v_6(t,z) = \frac{1}{36}(t^{10} - 27t^8z^2 + 42t^6z^4 + 42t^4z^6 - 27t^2z^8 + z^{10}).$$

Now let u be a solution of (a),(b) and v=v(y). Chose $y_1,..,y_8$, so that the right side of (c) is minimized:

Minimize $y_{n+1} + \frac{1}{2} y_{n+2}$

under the restrictions (P_S)

$\|(u-v)(t,z)\|$	$\leq y_{n+1}$	$\forall (t,z)\in \partial S$	inequality I	(u=0 and Δu=0 on ∂S)
$\|(\Delta u-\Delta v)(t,z)\|$	$\leq y_{n+2}$	$\forall (t,z)\in \partial S$	inequality II	

On symmetry reasons it is enough, to consider in (P_S) only those $(t,z)\in\partial S$ with t=1 and $z\in[0,1]$. Therefore the problem is only one dimensional.

Taking as grid measure h =1/100 and using Charnes M-method to get a starting basis solution we got the following results:

	y	positive dual variables in z =	
1	0.08855	1	(inequality I)
2	-0.38410	0	(inequality I)
3	-0.01885	1	(inequality I)
4	0.05162	0.61	(inequality I)
5	0.00034	0	(inequality II)
6	-0.00453	0.44	(inequality I)
7	0.00008	0.5	(inequality II)
8	0.00368	0.86	(inequality II)

The continuous problem seems to be regular.
CPU time for computing the dates 0.24 sec.

simplex method : 16 iterations 1.21 sec.

multiple exchange method : 6 multiple exchange steps (=19 single exchange steps for solving (15)) 0.68 sec

saving of CPU time : 44%

EXAMPLE 2 (complex approximation), compare OPFER [79]

The complex identity z is to be approximated by a complex polynomial $\sum_{i=2}^{n} y_i z^i$, $y \in C$.

Such questions are of interest in the theory of conformal mappings. If we use in the complex plane the norm $\|z\|_C = \max\{|\mathrm{real}(z)|, |\mathrm{imag}(z)|\}$ it can be shown that the image of the optimal error curve $z + \sum_{i=2}^{n} y_i z^i$ on S is a square for $n \to \infty$.

On symmetry reasons and since all functions are analytic it is sufficient to solve the following problem with real parameters:

$$\underset{y_2,\ldots,y_n}{\text{Minimize}} \left\| z + \sum_{j=2}^{n} y_j z^{4j-3} \right\|_C \quad \text{with } z \in \partial\tilde{S} = \{ z \mid z = e^{is}, \ s \in [0, \pi/4] \}$$

$$\Longleftrightarrow$$

Minimize y_{n+1} under the restrictions

$$\left.\begin{array}{lll} \text{I} & \sum_{j=2}^{n} y_j \,\mathrm{real}(z^{4j-3}) + y_{n+1} & \geq \mathrm{real}(z) \\ \text{II} & -\sum_{j=2}^{n} y_j \,\mathrm{real}(z^{4j-3}) + y_{n+1} & \geq -\mathrm{real}(z) \end{array}\right\} \quad \forall z \in \partial\tilde{S} = \{ z \mid z = e^{is}, s \in [0, \pi/4] \}$$

$$\left.\begin{array}{llll} \text{III} & \sum_{j=2}^{n} y_j \,\text{imag}(z^{4j-3}) & + \; y_{n+1} & \geq \text{imag}(z) \\ \text{IV} & -\sum_{j=2}^{n} y_j \,\text{imag}(z^{4j-3}) & + \; y_{n+1} & \geq -\text{imag}(z) \end{array}\right\} \quad \forall z \in \partial\check{S} = \{z \mid z = e^{is},\ s \in [0, \pi/4]\}$$

Taking 101 grid points and n = 7 we got the following results:

	y	positive dual variables in s =	
2	0.09471	0	
3	-0.03534	0.63	
4	0.01723	0.64	
5	-0.00901	0.416	each time in inequality I
6	0.00447	0.424	
7	-0.00165	0.20	
8	0.92959 (optimal value)	0.21	

We suggest that the continuous problem is singular with 4 positive dual variables.

CPU time for computing the dates 0.7 sec

simplex method	25 iterations	1.60 sec
multiple exchange method :	8 (=24) iterations	0.83 sec
saving CPU time :	48% (M-method)	

<u>EXAMPLE 3</u> (onesided L_1 approximation)

$$\text{Minimize } \left\| \sum_{i=1}^{n} y_i s^{i-1} - \tan(s) \right\|_{L_1} = \int_0^1 \left| \sum_{i=1}^{n} y_i s^{i-1} - \tan(s) \right| ds$$

under the restrictions

$$\sum_{i=1}^{n} y_i s^{i-1} \geq \tan(s) \qquad \forall s \in S = [0,1] .$$

Equivalently we have to solve the following problem:

$$\text{Minimize } \sum_{i=1}^{n} y_i c_i \qquad (\text{with } c_i = \int_0^1 s^{i-1}\, ds)$$

under the restrictions

$$\sum_{i=1}^{n} y_i s^{i-1} \geq \tan(s) \quad \forall s \in S$$

With 101 grid points and $n = 11$ we got the following results:

	y	positive dual variables in s =
1	0.000002	0.03
2	0.9999	0.04
3	0.0043	0.19
4	0.2794	0.20
5	0.3703	0.43
6	-1.3859	0.44
7	3.8931	0.69
8	-6.2487	0.70
9	6.2905	0.90
10	-3.5399	0.91
11	0.8944	1

optimal value 0.615627

The continuous problem is singular with 6 positive dual variables.
CPU time for computing the dates: 0.17 sec.

<u>2 phase method</u> (first phase single exchange 13 iterations 0.95 sec.)

simplex method	: 17 iterations	0.61 sec	(only second phase)
multiple exchange method	: 4 (=16) iterations	0.42 sec	"
saving CPU time	: 30%		

The problem is very instable, we got the above results with repartitions of the basis matrix in every iteration step using total pivot search.

The following **examples we solved infinitely** (two dimensional linear Chebyshev-approximation)

EXAMPLE 4 compare WATSON [75] and ANDREASSEN/WATSON [76]

$$\underset{y_1,\dots,y_6}{\text{Minimize}} \left\| e^{-t^2-z} - y_1 - y_2 t - y_3 z - y_4(2t^2-1) - y_5 tz - y_6(2z^2-1) \right\|$$

$S = [0,1] \times [0,1]$.

We got an optimal solution shown in the table of the next side. The problem is singular with 6 positive dual variables. A starting solution for the infinite programm was found on a grid of $9 \times 9 = 81$ points, therefore 0.75 seconds were spent.

simplex method	: 25 iterations	3.74 sec.	(infinite phase only)
multiple exchange method	: 11 (=28) iterations	2.35 sec.	"
saving CPU time	: 38%		

	exact y after ANDREASSEN/WATSON	y	positive dual variables in
1	0.98577	0.98577	(0.8356 ,0)
2	-0.34797	-0.34796	(0 ,1)
3	-0.90271	-0.90271	(0.67694,1)
4	-0.14462	-0.14462	(0.67686,1)
5	0.42457	0.42457	(1,0.6206)
6	0.11293	0.11293	(0.2721 ,0)
7	0.027274796 (optimal value)	0.027274796	(0,0.2177)

the simplex method delivered an about $2 \cdot 10^{-5}$ bader solution than the multiple exchange method

EXAMPLE 5 compare WATSON [75]

The search is for $(y_1, \ldots, y_9)^T$ to minimize

$$\|(t + 2z + 4)^{1/2} - y_1 - y_2 t - y_3 z - y_4 tz - y_5 t^2 z - y_6 t^2 z^2 - y_7 tz^2 - y_8 z^2 - y_9 t^2\|$$

$S = [-1,1] \times [-1,1]$.

With the multiple exchange method we got the following solution

	y	positive dual variables in
1	2.0086	(-1,-1)
2	0.2513	(-1,1)
3	0.5123	(-1,-0.6230)
4	-0.0698	(-1,0.3839)
5	0.0245	
6	-0.0148	all other dual variables were less 5 10^{-12}
7	0.0336	
8	-0.0707	
9	-0.0201	
10	0.01140060 (optimal value)	

we suggest that the problem has a singular solution with 4 positive dual variables

We needed 1.36 seconds to determine a starting basis solution on a grid of 9x9 = 81 points. The optimal value is identical with the value computed by WATSON.

simplex method	: 21 iterations	5.60 sec	(infinite phase only)
multiple exchange method	: 9 (=23) iterations	2.92 sec	"
saving of CPU time	: 47%		

REFERENCES

1) Andreassen, D.O. Watson, G.A. : Linear Chebyshev approximation without Chebyshev sets, BIT 16 (1976), 349-362 .

2) Bartels, R.H. Golub, G.H. : Stable numerical methods for obtaining the Chebyshev solution to an overdetermined system of equations, CACM, Vol. 11 , 6, (1968).

3) Bartels, R.H. Golub, G.H. : The simplex method of linear programming using LU-decompositions, CACM 12, (1969),266-268.

4) Bredendiek, E. Collatz, L. : Simultanapproximation bei Randwertaufgaben. Numerische Methoden der Approximationstheorie, Bd. 3, ISNM 30, 147-174, Birkhäuser-Verlag, Basel, Stuttgart 1976.

5) Daniel, J.W. Gragg, W.B. Kaufmann, L. Stewart, G.W. : Stable algorithms for updating the Gram-Schmidt QR factorization , Math. Comp. 30 (1976), 772-795.

6) Dantzig, G.B. : Minimization of a linear function subject to linear inequalities, in Koopmans: Activity analysis of production and allocation, John Wiley and Sons, New York (1951).

7) Gill, P.E. Golub, G.H. Murray,W. Saunders, M.A. : Methods for modifying matrix factorizations, Math. Comp. 28, (1974), 505-535.

8) Gill, P.E. Murray, W. : A numerical stable form of the simplex algorithm, Linear algebra and applications 7 (1973), 99-138.

9) Glashoff, K. Gustafson, S-A. : Einführung in die lineare Optimierung, Wissenschaftliche Buchgesellschaft, Darmstadt (1978).

10) Hoffmann, K.-H. Klostermair, A. : A semi-infinite linear Programming procedure and applications to approximation problems in optimal control. Appr. Theory II, Proc. int. Symp., Austin (1976),379-389.

11) Judin, D.B. Golstein, E.G. : Lineare Optimierung I, Akademieverlag, Berlin (1968).

12) Klostermair, A. : Austauschalgorithmen zur Lösung konvexer Optimierungsaufgaben, Diplomarbeit, München (1974).

13) Opfer, G. : New Extremal Properties for constructing Conformal Mappings, submitted to Numerische Mathematik (1979).

14) Orchard-Hays, W. : Advanced linear programming computing techniques, Mc. Graw Hill, New York (1968)

15) Stoer, J. : Einführung in die numerische Mathematik, 2.ed., Springer Verlag, Berlin-Heidelberg-New York (1976).

16) Watson, G.A. : A multiple exchange algorithm for multivariate Chebyshev Approximation, SIAM J. Num. Anal. 12 (1975), 46-52.

17) Roleff, K. : Ein stabiles Mehrfachaustauschverfahren für lineare semi-infinite Optimierungsprobleme, Dissertation,Hamburg,1979.

ON QUADRATICALLY CONVERGENT METHODS FOR SEMI-INFINITE PROGRAMMING

R. Hettich, W. van Honstede
Institut für Angewandte Mathematik
Universität Bonn
Wegelerstr. 6
5300 Bonn, Germany

Abstract. A class of methods for solving general nonlinear semi-infinite programming problems is considered which may be shown to converge superlinearly to a solution, if for this solution a sufficient second order optimality condition holds. An important feature of all these methods is that they are related to the treatment of a finite programming problem. In the last two sections generalizations of "approximation methods" from nonlinear programming to the semi-infinite case are considered.

1. Introduction. In this contribution we consider a class of numerical methods for solving the following semi-infinite problem:

Problem (O). Given an open set $X_O \subset \mathbb{R}^n$, a compact set $Y \subset \mathbb{R}^m$, and twice continuously differentiable functions $F\colon X_O \to \mathbb{R}$, $f\colon X_O \times Y \to \mathbb{R}$. Minimize $F(x)$ subject to $x \in X$, where the feasible region X is given by

$$X = \{x \mid f(x,y) \leq 0,\ y \in Y\}. \tag{1.1}$$

Let $\hat{x} \in X$ be a given point, $\hat{Y} \subset Y$ the corresponding set of active points, i.e.

$$\hat{Y} = \{y \in Y \mid f(\hat{x},y) = 0\}. \tag{1.2}$$

Given the following Property (A), it is possible locally to reduce the problem to one with a finite number of constraints.

Property (A). $\hat{Y}$ is a finite set: $\hat{Y} = \{\hat{y}^1,\dots,\hat{y}^r\}$. There are neighborhoods $\hat{X}$, $\hat{Y}_i$ of $\hat{x}$, $\hat{y}^i$, $i = 1,\dots,r$, and continuously differentiable functions $y^i\colon \hat{X} \to Y \cap \hat{Y}_i$ such that

(i) $y^i(\hat{x}) = \hat{y}^i$, $i = 1,\dots,r$

(ii) For every $\tilde{x} \in \hat{X}$, $y^i(\tilde{x})$ is the only local maximum of $f(\tilde{x},y)$ in $Y \cap \hat{Y}_i$.

Now we consider the following finite optimization problems:

Problem (FO). Minimize $F(x)$ subject to $x \in \hat{X}$ and

$$\phi^i(x) := f(x,y^i(x)) = 0, \ i = 1,\dots,r. \tag{1.3}$$

Problem (FO'). Let $\bar{Y} \subset Y$ be a finite subset of Y. Minimize $F(x)$ subject to $x \in \hat{X}$, $\phi^i(x) \leq 0$, $i = 1,\dots,r$, and $f(x,y^l) \leq 0$ for $y^l \in \bar{Y}$.

Then we have

Theorem 1. If Property (A) holds, then

(a) $\hat{x} \in \hat{X}$ is (strictly) locally optimal for (O) if and only if it is (strictly) locally optimal for (FO').
(b) If $\hat{x}$ is (strictly) locally optimal for (O), then it is (strictly) locally optimal for (FO).

Proof. (a) Denote by X' the set of feasible points of (FO'). Then, the inequality constraints of (FO') being constraints for (O) as well, $\hat{X} \cap X \subsetneq X'$, showing that a (strictly) locally optimal point of (FO') is (strictly) locally optimal for (O).

To prove the converse, let $\hat{x}$ be locally optimal for (O). Suppose $\hat{x}$ is not locally optimal for (FO'). Then there is a sequence of points $x^i \in X'$ such that $\lim_{i\to\infty} x^i = \hat{x}$ and $F(x^i) < F(\hat{x})$ for all i. We are ready if we show that there is a number i_0 such that $x^i \in X$ for $i > i_0$.

Due to the definition of $\hat{Y}$, there are neighborhoods $\tilde{Y}_i \subset \hat{Y}_i$ such that $\sup\{f(\hat{x},y) \mid y \in Y \sim \bigcup_{i=1}^{r} \tilde{Y}_i\} = d < 0$. Continuity of f implies that there is a neighborhood $\tilde{X} \subset \hat{X}$ such that

$$\sup\{f(x,y) \mid y \in Y \sim \bigcup_{i=1}^{r} \tilde{Y}_i,\ x \in \tilde{X}\} < 0.$$

By Property (A), $x^i \in X'$ implies $f(x,y) \leq 0$ for $x \in \tilde{X}$, $y \in \tilde{Y}_i$, $i=1,\dots,r$. Therefore, $x^i \in X$ if $x^i \in \tilde{X} \cap X'$, the latter being the case for i sufficiently large.

For strictly locally optimal points the proof is analoguous.

(b) The proof follows the same line as that for the second part of (a).

This possibility of reducing the semi-infinite problem locally to a finite one will give rise to a whole class of methods. We emphasize that (FO) and (FO') are related to (O) much more closely than the problem obtained by replacing Y by some finite subset. This is illustrated by the following example (cf. [5]):

The problem of approximating the function $g(y) = 1 - y^2$ in $[-1,1]$ by a function from the set $\{a(x,y) = \frac{1}{2}x^2-2xy \mid x \in \mathbb{R}\} \subset C[-1,1]$ in the sense of Chebyshev is equivalent to minimizing $F(x,d) = d$ subject to $|g(y) - a(x,y)| \leq d,\ y \in [-1,1]$.

$x = 0$ is a locally best approximation and consequently $\binom{0}{1}$ is a locally optimal solution to the optimization problem.

We find $\hat{Y} = \{0\}$ and $y^i(x) = x$. Therefore, (FO) becomes:

Minimize $F(x,d) = d$ subject to $1 - x^2 - \frac{1}{2}x^2 + 2x^2 = 1 + \frac{1}{2}x^2 = d$ with unique solution $\binom{0}{1}$.

On the other hand, it is easily seen that there is no finite subset of $[-1,1]$ such that the corresponding finite problem has the same solution as the semi-infinite one.

Actually, it is this closer relationsship which makes methods based on (FO) or (FO') superior to those which are based on solving discretized problems (cf. [6] for a more detailed discussion).

In the next section we first give a condition on that we may assume (O) to have Property (A) and recall a sufficient condition for $\hat{x}$ to be locally strictly optimal for (O), which is basic for the proofs of convergence of the methods to be described in the sequel. Proofs and a more detailed discussion of optimality conditions may be found in [7]. Then, assuming this sufficient condition to hold in $\hat{x}$, we prove the regularity of a matrix appearing in the proofs of convergence lateron. Moreover, a Newton method originally proposed by Gustafson for the case of linear problems, is described in Section 2. In Section 3 an alternative to this method, based on (FO) is given and, in Section 4, some remarks on the use of (FO') are made. In Section 5 and Section 6 approximation methods (such as Wilson's method) are considered.

2. Second order optimality conditions; convergence of Newton's method.

From now on, we restrict ourselves to a more specific class of problems (O), where the compact set $Y \subset \mathbb{R}^m$ is assumed to be given by a finite number of inequalities:

Let $g^j: \mathbb{R}^m \to \mathbb{R}$, $j \in J$, $|J| < \infty$, be given, twice continuously differentiable functions. Then

$$Y = \{y | g^j(y) \leq 0, \ j \in J\}. \tag{2.1}$$

Furthermore, the following Property (A') is assumed to be given at the point $\hat{x} \in X$.

Property (A'). For $y \in Y$ let $J(y) = \{j \in J | g^j(y) = 0\}$.

(i) For every $y \in Y$ the gradients $g^j_y(y)$, $j \in J(y)$, are linearly independent.

(ii) Let $\hat{Y} = \{\hat{y}^1,\dots,\hat{y}^r\}$ be defined as before. Then, for $i = 1,\dots,r$, there are $\hat{w}^{ij} > 0$, $j \in J(\hat{y}^i)$ such that

$$f_y(\hat{x},\hat{y}^i) - \sum_{j \in J(\hat{y}^i)} \hat{w}^{ij} g^j_y(\hat{y}^i) = 0 \tag{2.2}$$

and such that the matrix

$$\hat{M}^i = f_{yy}(\hat{x},\hat{y}^i) - \sum_{j \in J(\hat{y}^i)} \hat{w}^{ij} g^j_{yy}(\hat{y}^i) \tag{2.3}$$

is negative definite on the subspace

$$\hat{S}^i = \{\zeta | \zeta^T g^j_y(\hat{y}^i) = 0, \ j \in J(\hat{y}^i)\} \subset \mathbb{R}^m. \tag{2.4}$$

We remark that (ii) in (A') is sufficient for the points $\hat{y}^i$ to be strict local maxima of $f(\hat{x},y)$ on Y (cf. [8] or [3]).

Theorem 2. Property (A') is sufficient for Property (A) to be given. Moreover, there are neighborhoods $\hat{W}_{ij}$ of $\hat{w}^{ij}$ and continuously differentiable functions $w^{ij}: \hat{U} \to \hat{W}_{ij}$ such that $w^{ij}(\hat{x}) = \hat{w}^{ij}$ and such that for all $x \in \hat{U}$

$$f_y(x,y^i(x)) - \sum_{j \in J(\hat{y}^i)} w^{ij}(x) g^j_y(y^i(x)) = 0 \tag{2.5}$$

$$g^j(y^i(x)) = 0, \ j \in J(\hat{y}^i). \tag{2.6}$$

Let w^i be the vector with elements w^{ij}, $j \in J(\hat{y}^i)$, and $\hat{G}^i$ the matrix with columus $g^j_y(\hat{y}^i)$, $j \in J(\hat{y}^i)$. Then the matrix

$$\hat{A}^i = \begin{pmatrix} \hat{M}^i & \hat{G}^i \\ (\hat{G}^i)^T & 0 \end{pmatrix} \tag{2.7}$$

is nonsingular and the Hessian matrices $y_x^i(\hat{x})$, $w_x^i(\hat{x})$ are uniquely determined by

$$\hat{A}^i \begin{pmatrix} y_x^i(\hat{x}) \\ w_x^i(\hat{x}) \end{pmatrix} = \begin{pmatrix} -f_{yx}(\hat{x},\hat{y}^i) \\ 0 \end{pmatrix} \tag{2.8}$$

Proofs of this and the two following theorems may be found in [7].

<u>Theorem 3</u>. Suppose Property (A') to be given. Then, if $\hat{x}$ is optimal, for every $\xi \in K$,

$$K = \{\xi \mid \xi^T F_x(\hat{x}) \leq 0,\ \xi^T f_x(\hat{x},\hat{y}^i) \leq 0,\ i = 1,\dots,r\} \tag{2.9}$$

there are $\hat{u}^0 \geq 0$, $\hat{u}^i \geq 0$, not all equal to zero, such that

$$\hat{u}^0 F_x(\hat{x}) + \sum_{i=1}^{r} \hat{u}^i f_x(\hat{x},\hat{y}^i) = 0 \tag{2.10}$$

and such that

$$q(\hat{u}^0,\hat{u}^i,\xi) = \xi^T(\hat{u}^0 F_{xx}(\hat{x}) + \sum_{i=1}^{r} \hat{u}^i f_{xx}(\hat{x},\hat{y}^i))\xi - \sum_{i=1}^{r} \hat{u}^i (\hat{\mu}^i(\xi))^T \hat{M}^i \hat{\mu}^i(\xi) \geq 0 \tag{2.11}$$

where $\hat{\mu}^i(\xi)$ are the unique solutions of

$$\hat{A}^i \begin{pmatrix} \hat{\mu}^i(\xi) \\ \rho(\xi) \end{pmatrix} = \begin{pmatrix} -f_{yx}(\hat{x},\hat{y}^i)\xi \\ 0 \end{pmatrix} . \tag{2.12}$$

<u>Theorem 4</u>. Assume Property (A') to be given. Then, if for every $\xi \in K$ there exist $\hat{u}^0 \geq 0$, $\hat{u}^i \geq 0$, $i = 1,\dots,r$, such that (2.10) holds and, if $\xi \neq 0$, $q(\hat{u}^0,\hat{u}^i,\xi) > 0$, then $\hat{x}$ is strictly locally optimal.

In the sequel, the following Property (B) will play a central role, being a combination of Property (A') and a stronger version of the sufficient condition in Theorem 4.

<u>Property (B)</u>. Property (A') is given. The vectors $f_x(\hat{x},\hat{y}^i)$, $i=1,\dots,r$, are linearly independent and the sufficient condition of Theorem 4 holds with $\hat{u}^0 = 1$, $\hat{u}^i > 0$, $i = 1,\dots,r$.

Note that the linear independence of $f_x(\hat{x},\hat{y}^i)$ implies $\hat{u}^0 \neq 0$ in (2.10). Moreover, choosing $\hat{u}^0 = 1$, the $\hat{u}^i$, $i = 1,\dots,r$, are uniquely determined.

A well-known method for solving (O), originally proposed by Gustafson [4] for linear problems, consists in applying Newton's method to the nonlinear system of equations

$$F_x(x) + \sum_{i=1}^{r} u^i f_x(x,y^i) = 0 \qquad (2.13)$$

$$f(x,y^i) = 0, \; i = 1,\dots,r \qquad (2.14)$$

$$f_y(x,y^i) - \sum_{j\in J(\hat{y}^i)} w^{ij} g_y^j(y^i) = 0, \; i = 1,\dots,r \qquad (2.15)$$

$$g^j(y^i) = 0, \; j \in J(\hat{y}^i), \; i = 1,\dots,r \qquad (2.16)$$

with unknowns x, y^i, u^i, w^{ij}. (2.10), (2.2) show that $\hat{x}$, $\hat{y}^i$, $\hat{u}^i$, and $\hat{w}^{ij}$ are solutions of (2.13)-(2.16).

Collecting the left hand sides of (2.13)-(2.16) in one function Φ, we have a system

$$\Phi(x,u^i,y^i,w^{ij}) = 0. \qquad (2.17)$$

It is well-known that, if the Jacobian $\Phi_{(x,u^i,y^i,w^{ij})}$ is nonsingular, Newton's method will be locally convergent, the convergence being superlinear and, under mild additional differentiability assumptions, even quadratic.

Theorem 5. If Property (B) holds, then the Jacobian $\Phi_{(x,u^i,y^i,w^{ij})}$ of system (2.17) is nonsingular.

We omit the lengthy but elementary proof which may be found in [5] for the special case of Chebyshev-approximation problems.

A disadvantage of the above method is that convergence may be lost if for instance for one of the points $\hat{y}^i$ the matrix $\hat{M}^i$ (2.3) happens to be singular.
In the method presented in the following section the search for y^i is done independently, therefore this version seems to be more useful in practice.

3. Newton's method based on problem (FO).

Instead of (2.13)-(2.16) we may consider the corresponding system of nonlinear equations for problem (FO). Necessary for $\hat{x}$ to be optimal for (FO) is the existence of $\hat{u}^i$ such that

$$F_x(\hat{x}) + \sum_{i=1}^{r} \hat{u}^i f_x(\hat{x}, y^i(\hat{x})) = 0.$$

Together with the equality constraints, this leads to the following equations for the unknowns x, u^i:

$$F_x(x) + \sum_{i=1}^{r} u^i f_x(x, y^i(x)) = 0 \tag{3.1}$$

$$f(x, y^i(x)) = 0, \; i = 1,\dots,r \tag{3.2}$$

Writing F_{xx}, f^i_{xy} etc. instead of $F_{xx}(\hat{x})$, $f_{xy}(\hat{x}, y^i(\hat{x}))$ etc. the Jacobian of (3.1), (3.2) for $x = \hat{x}$, $u^i = \hat{u}^i$ becomes

$$J(\hat{x},\hat{u}) = \left(\begin{array}{c|c} F_{xx} + \sum_{i=1}^{r} \hat{u}^i \{f^i_{xx} + f^i_{xy} y^i_x\} & f^i_x \;\dots\; f^r_x \\ \hline (f^1_x)^T + (f^1_y)^T y^1_x & \\ \vdots & 0 \\ (f^r_x)^T + (f^r_y)^T y^r_x & \end{array}\right) \tag{3.3}$$

From $f(x, y^i(x)) \equiv 0$ it follows that the terms $(f^i_y)^T y^i_x$ vanish. To compute $f^i_{xy} y^i_x$ we multiply (2.8) from the left by $(f^i_{xy}|0)(\hat{A}^i)^{-1}$. This yields

$$f^i_{xy} y^i_x = (-f^i_{xy}|0)(\hat{A}^i)^{-1} \left(\frac{f^i_{yx}}{0}\right) \tag{3.4}$$

<u>Theorem 6</u>. If Property (B) is given, then the Jacobian (3.3) of (3.1), (3.2) is nonsingular.

<u>Proof</u>. Suppose J is singular. Then there is a vector $\binom{\xi}{\eta} \neq 0$, $\xi \in \mathbb{R}^n$, $\eta \in \mathbb{R}^r$, such that $J\binom{\xi}{\eta} = 0$.

The linear independence of the vectors $f^1_x, \dots, f^r_x$ implies $\eta = 0$ if $\xi = 0$. Therefore, assume $\xi \neq 0$. Then $(f^i_x)^T \xi = 0$, $i = 1,\dots,r$, and, by (2.10), $F^T_x \xi = 0$, i.e., $\xi \in K$ (cf. (2.9)).

The first n equations of $J\binom{\xi}{\eta} = 0$ are

$$(F_{xx} + \sum_{i=1}^{r} \hat{u}^i \{f^i_{xx} + f^i_{xy} y^i_x\})\xi + \sum_{i=1}^{r} \eta_i f^i_x = 0.$$

Multiplication by ξ^T from the left gives

$$\xi^T(F_{xx} + \sum_{i=1}^{r} \hat{u}^i f^i_{xx})\xi + \sum_{i=1}^{r} \hat{u}^i \xi^T f^i_{xy} y^i_x \xi = 0 \qquad (3.5)$$

(2.12), (3.4) imply

$$\xi^T f^i_{xy} y^i_x \xi = \xi^T(-f^i_{xy}|0)(\hat{A}^i)^{-1}\left(\frac{f^i_{yx}}{0}\right)\xi = -(\hat{\mu}^i(\xi))^T \hat{M}^i \hat{\mu}^i(\xi) \quad .$$

Therefore, (3.5) becomes

$$q(1,\hat{u}^i,\xi) = 0 \quad ,$$

a contradiction to Property (B), finishing the proof.

This gives rise to, for instance, the following algorithm:

Start: Given some approximation x^0 to $\hat{x}$, compute local maxima $y^{0,1},\ldots$ $\ldots,y^{0,r}$ of $f(x^0,y)$ on Y. Suppose that part (ii) of Property (A') holds with x^0, $y^{0,i}$, w_0^{ij} instead of $\hat{x}$, $\hat{y}^i$, $\hat{w}^{ij}$. Determine best least squares approximations $u^{0,i}$ to $\hat{u}^i$ such that (3.1) holds best possible.

Step i ($i \geq 1$). (1) Compute new approximations x^i, $u^{i,j}$, $j = 1,\ldots,r$, by performing one step of Newton's method applied to (3.1), (3.2), i. e.

$$J(x^{i-1},u^{i-1})\begin{pmatrix}\Delta x^i\\ \Delta u^i\end{pmatrix} = -\begin{pmatrix} F_x(x^{i-1}) + \sum_{j=1}^{r} u^{i-1,j} f_x(x^{i-1},y^{i-1,j}) \\ f(x^{i-1},y^{i-1,j}),\ j = 1,\ldots,r \\ \vdots \end{pmatrix}$$

$$x^i = x^{i-1} + \Delta x^i,\ u^i = u^{i-1} + \Delta u^i,$$

$J(x^{i-1},u^{i-1})$ according to (3.3).

(2) Compute $y^{i,j}$, w_i^{jl}, $l \in J(\hat{y}^j)$, $j = 1,\ldots,r$, as local maxima of $f(x^i,y)$ on Y and corresponding parameters to solve (2.2).

The above algorithm as well as that of Section 2 has the disadvantage that the number r of points $\hat{y}^i$ should be known and for $i = 1,\ldots,r$, the index sets $J(\hat{y}^i)$. Yet this is less serious in the above algorithm in so far as in each step the $y^{i,j}$ are computed separately allowing a better control over possible errors in the initial choices of r, $J(\hat{y}^i)$. Asymptotically the two methods become equivalent, if Property (B) is given.

For a more detailed discussion cf. [6].

4. Methods based on problem (FO').

A very promising, but up to now only occasionally treated way seems to us to consider problem (FO') rather than (FO). The following advantages are obvious:

- Treating (FO), no control will be made on the signs of the $u^{i,j}$, whereas this would be the case in a very natural way if (FO') is considered.
- By not only taking into account variable points $y^i(x)$ corresponding to maxima of $f(x^i,y)$ for the actual approximation but moreover some net $\overline{Y}$ of fixed points, the set of active points can be adapted very easily in the course of the computational process. This implies that the assumptions on the starting point can be weakened considerably.

An example of such an algorithm may be found in [9]. Naturally, a lot of methods from (finite) nonlinear programming can be used to solve (FO').

5. Approximation methods.

Let $\hat{x} \in X$, locally optimal for (O), be given such that Property (B) holds with $\hat{u} = \begin{pmatrix}\hat{u}^1\\ \vdots\\ \hat{u}^r\end{pmatrix} \in \mathbb{R}^r$, $\hat{y} = \begin{pmatrix}\hat{y}^1\\ \vdots\\ \hat{y}^r\end{pmatrix} \in \mathbb{R}^{rm}$, and $\hat{w} = \begin{pmatrix}\hat{w}^1\\ \vdots\\ \hat{w}^r\end{pmatrix}$, where for $i = 1,\ldots,r$, $\hat{w}^i = \begin{pmatrix}\vdots\\ \hat{w}^{ij}\\ \vdots\end{pmatrix}_{j\in J(\hat{y}^i)} \in \mathbb{R}^{|J(\hat{y}^i)|}$.

A method will be called an approximation method for (O) if it proceeds as follows:

Suppose, for each $z \in \hat{S}$, $\hat{S}$ some neighborhood of $\hat{z}$, in a way to be specified lateron, a semi-infinite problem (O_z) is defined.

Algorithm (AG):

Step 1. Choose $z^o \in \hat{S}$.
Step 2. Given $z^{i-1} \in \hat{S}$. Determine z^i as a solution of $(O_{z^{i-1}})$. If the solution is not unique take the one closest in norm to z^{i-1}.
Step 3. Test for convergence. Stop or go to Step 2. (Here it is assumed that $z^i \in \hat{S}$).

That means, in each step the given problem is "approximated" in some sense by another one which naturally should be easier to solve than the given one.

An obvious way is to take as (O_{z^i}) the problem obtained by linearizing the functions F, f with respect to x in $x = x^i$. This "linear approximation method" can be proved to converge quadratically to $\hat{z}$ if $\hat{x}$ is locally strongly unique, a rather restrictive assumption (cf. [1] for Chebyshev approximation problems).

In this section we will define a class of methods locally convergent to a $\hat{z}$ with Property (B), an assumption considerably weaker than strong uniqueness. In exchange, problem (O_z) is no longer linear but, typically, linearly restricted only.

Our analysis follows the same line as that in [11] for finite problems. First, a sensitivity result is obtained by means of which convergence can be established easily. We remark that our sensitivity result is slightly less general than that given in [11] inasmuch as - to save space - we immediately impose Property (B) (similar to [2] for the finite case) and, moreover, consider perturbations of $\hat{z}$ explicitely.

Now, let us define (O_z) more specifically.

<u>Problem (O_{z^i})</u>. Let $\hat{S}$ be an open neighborhood of $\hat{z}$, $\tilde{F}(x,z)$, $\tilde{f}(x,y,z)$ be twice continuously differentiable and defined on $X_O \times \hat{S}$ and $X_O \times Y \times \hat{S}$ resp. Given $z^i \in \hat{S}$, minimize $\tilde{F}(x,z^i)$ subject to $x \in X_{z^i}$, X_{z^i} given by

$$X_{z^i} = \{x \mid \tilde{f}(x,y,z^i) \leq 0,\ y \in Y\} \quad . \tag{5.1}$$

Let $\tilde{\Phi}(z,z^i)$ be defined accordingly to (2.17) with $\tilde{F}$, $\tilde{f}$ instead of F, f ($r, J(\hat{y}^i)$ are fixed!). Then, $\tilde{F}$, $\tilde{f}$ are supposed to be given such that for every $z \in \hat{S}$ we have

$$\tilde{\Phi}(z,z) = \Phi(z),\ \tilde{\Phi}^1_z(z,z) = \Phi_z(z) \tag{5.2}$$

where superscript 1 indicates that the derivative is to be taken with respect to the first argument z.

We emphazise that in the sequel solving (O_{z^i}) means that an optimal point x is determined as well as z such that $\tilde{\Phi}(z,z^i) = 0$.

The following lemma is obvious from the differentiability assumptions imposed on $\tilde{F}$, $\tilde{f}$.

<u>Lemma</u>. For every compact subset $\tilde{S} \subset \hat{S}$ there exist $\lambda \geq 1$, $\alpha \in \mathbb{R}$ such that for all z^1, $z^O \in \tilde{S}$

$$\| \tilde{\Phi}(z^1,z^1) - \tilde{\Phi}(z^1,z^0) \| \le \alpha \| z^1 - z^0 \|^{\lambda}. \tag{5.3}$$

Here and in the sequel $\| \cdot \|$ is a fixed vectornorm.

Theorem 7. Suppose Property (B) holds. Let $\beta = \| (\Phi_z(\hat{z}))^{-1} \|$ (Φ_z^{-1} exists due to Theorem 5) and $0 < \delta < \beta^{-1}$ a given number. Then there exist neighborhoods $\hat{S}_1, \hat{S}_2 \subset \hat{S}$ of $\hat{z}$ and $Z: \hat{S}_2 \to \hat{S}_1$, continuously differentiable, such that $Z(\hat{z}) = \hat{z}$ and such that for every fixed $z \in \hat{S}_2$ we have:

(i) $Z(z)$ is the unique solution of $\tilde{\Phi}(\cdot,z) = 0$ in $\hat{S}_1$.

(ii) $Z_z(z) = -[\tilde{\Phi}_z^1(Z(z),z)]^{-1}\tilde{\Phi}_z^2(Z(z),z)$ (5.4)

(iii) $Z(z)$ is the unique solution of (O_z) in $\hat{S}_1$ and Property (B) holds in $Z(z)$. Note that by (i), r, $J(y^i(z))$ are the same as r and $J(\hat{y}^i)$ for $\hat{z}$ and (O).

(iv) $\| Z(Z(z)) - Z(z) \| \le \alpha\beta/(1-\delta\beta) \| Z(z) - z \|^{\lambda}.$ (5.5)

Proof. We first show that $\hat{z}$ has Property (B) with respect to $(O_{\hat{z}})$.

By (5.2) we have $\Phi_x(\hat{z}) = \tilde{\Phi}_x^1(\hat{z},\hat{z})$, thus (cf. (2.13)-(2.17))

$$F_{xx}(\hat{x}) + \sum_{i=1}^{r} \hat{u}_i f_{xx}(\hat{x},\hat{y}_i) = \tilde{F}_{xx}(\hat{x},\hat{z}) + \sum_{i=1}^{r} \hat{u}_i \tilde{f}_{xx}(\hat{x},\hat{y}^i,\hat{z}) \tag{5.6}$$

$$\left.\begin{aligned} f_x(\hat{x},\hat{y}^i) &= \tilde{f}_x(\hat{x},\hat{y}^i,\hat{z}) \\ f_{yx}(\hat{x},\hat{y}^i) &= \tilde{f}_{yx}(\hat{x},\hat{y}^i,\hat{z}) \end{aligned}\right. \quad i = 1,\dots,r \tag{5.7}$$

(Recall that Y is the same for (O) and (O_z)!).

Similarly $\Phi_y(\hat{z}) = \tilde{\Phi}_y^1(\hat{z},\hat{z})$ gives

$$\left.\begin{aligned} f_y(\hat{x},\hat{y}^i) &= \tilde{f}_y(\hat{x},\hat{y}^i,\hat{z}) \\ f_{yy}(\hat{x},\hat{y}^i) &= \tilde{f}_{yy}(\hat{x},\hat{y}^i,\hat{z}) \end{aligned}\right\} \quad i = 1,\dots,r \tag{5.8}$$

Finally, using $\tilde{\Phi}(\hat{z},\hat{z}) = \Phi(\hat{z})$ and (5.7) we have

$$F_x(\hat{x}) = \tilde{F}_x(\hat{x},\hat{z}). \tag{5.9}$$

From (5.6)-(5.9) one immediately concludes that $\hat{z}$ has Property (B) with respect to $(O_{\hat{z}})$ if and only if it has Property (B) with respect to (O).

Especially we have proved $\tilde{\Phi}(\hat{z},\hat{z}) = 0$. By Theorem 5 we have $\tilde{\Phi}_z^1(\hat{z},\hat{z})$ is regular. Therefore, the implicit function theorem yields the existence of neighborhoods $\tilde{S}_1, \tilde{S}_2 \subset \hat{S}$ of $\hat{z}$ and a continuously differentiable

$Z: \tilde{S}_2 \to \tilde{S}_1$ such that $Z(\hat{z}) = \hat{z}$ and (i) holds for $\hat{S}_1 = \tilde{S}_1$. Moreover, we may assume that $\tilde{\Phi}^1_z(z^1,z^2)$ is regular in $\tilde{S}_1 \times \tilde{S}_2$. Therefore, applying the implicit function theorem again, we find for every $z \in \tilde{S}_2$ that (5.4) holds.

According to the proof of the implicit function theorem in [10, 5.2.4.] we may further assume that $\tilde{S}_1$, $\tilde{S}_2$ are choosen such that $Z(z) \in \tilde{S}_2$ implies

$$\| Z(Z(z))-Z(z) \| \leq \beta(1-\delta\beta)^{-1} \| \tilde{\Phi}(Z(z),Z(z))-\tilde{\Phi}(Z(z),z) \| \tag{5.10}$$

Therefore, using (5.3), (iv) is shown for $z \in \hat{S}_2$, $\hat{S}_2$ such that $Z(z) \in \tilde{S}_2$.

It remains to prove that for z in a possibly smaller neighborhood in Z(z) Property (B) holds with respect to (O_z). Using the fact that it is true for $z = \hat{z}$ this proof is straightforward due to the twice continuous differentiability of $\tilde{F}$, $\tilde{f}$.

Using Theorem 7 in a way completely analoguous to [11] the following is easily proved (δ of Theorem 7 chosen $\beta^{-1}(1-\alpha\beta)/(1+\alpha\beta)$ for part (a) and $\delta = \beta^{-1}/2$ for part (b)).

Theorem 8. (a) If (5.3) holds with $\lambda = 1$ and α such that $\alpha\beta < 1$, then there is a neighborhood $\tilde{S} \subset \hat{S}$ such that for $z^0 \in \tilde{S}$ the sequence $\{z^k\}$ generated by (AG) exists and converges to $\hat{z}$, the convergence being linear in the sense that

$$\| \hat{z} - z^k \| \leq \gamma^k \cdot c \tag{5.11}$$

with $\gamma = \frac{1}{2}(1+\alpha\beta) < 1$ and $c = \frac{2}{1-\alpha\beta} \|z^1-z^0\|$.

(b) If (5.3) holds with $\lambda > 1$, z^k as in (a) exists and converges to $\hat{z}$, the convergence being superlinear in the sense that

$$\|\hat{z} - z^k\| \leq \sigma^{\lambda^k} \mu \tag{5.12}$$

with $\sigma = (2\alpha\beta)^{1/(1-\lambda)} \|z^1-z^0\| < 1$, $\mu = (2\alpha\beta)^{1/(1-\lambda)}(1-\sigma^{\lambda-1})^{-1}$.

Therefore, each choice of (O_z) such that (5.2) and (5.3) with either $\lambda > 1$ or $\alpha\beta < 1$ holds, leads to a locally convergent algorithm for problems where in the optimal point $\hat{z}$ Property (B) is given. In the next section we give an example for such an algorithm (see also [9]).

6. Generalization of Wilson's method to the semi-infinite case.

In [11] several special approximation methods are given for the finite case. One of them, given first by Wilson [12], we now will generalize to the semi-infinite case. For introduction we first formulate the algorithm obtained by applying Wilson's method to the finite problem (FO'):

Algorithm WI. Given z^k, a problem (FO_k') is defined as follows:

Problem (FO_k'). Minimize the function

$$F^*(x,z^k) = F(x^k) + (x-x^k)^T F_x(x^k) + \frac{1}{2}(x-x^k)^T L_{xx}(z^k)(x-x^k)$$

subject to the linear constraints

$$f^*(x,y^i(x^k),z^k) = f(x^k,y^i(x^k)) + (x-x^k)^T f_x(x^k,y^i(x^k)) \leq 0,$$

$$i = 1,\dots,r$$

$$f^*(x,y^l,z^k) = f(x^k,y^l) + (x-x^k)^T f_x(x^k,y^l) \leq 0, \; y^l \in \bar{Y} \quad .$$

Here, L is the Lagrangian

$$L(z) = F(x) + \sum_{i=1}^{r} u^i f(x,y^i(x)) \quad .$$

Using (2.2), (2.8), this gives

$$L_{xx}(z) = F_{xx}(x) + \sum_{i=1}^{r} u^i \{f_{xx}(x,y^i(x)) + f_{xy}(x,y^i(x))y_x^i(x)\}.$$

Let z^k be an approximation of the optimal point $\hat{z}$ such that y^{ik} are local maxima of $f(x^k,y)$ and part (ii) of Property (A') holds for (x^k,y^k,w^k). Then (2.15) and (2.16) hold and one step of Newton's method applied to the linear system of equations

$$F_x^*(x,z^k) + \sum_{i=1}^{r} u^i f_x^*(x,y^i(x^k),z^k) = 0 \qquad (6.1)$$

$$f^*(x,y^i(x^k),z^k) = 0, \; i = 1,\dots,r \qquad (6.2)$$

(it may be proved that for z^k sufficiently close to $\hat{z}$ this means that FO_k' is solved) is equivalent to one step of the Newton method of Section 3. Therefore, the superlinear (and under mild additional differentiability assumptions quadratic) convergence of Newton's method implies the same for algorithm WI.

For semi-infinite problems the method above can be generalized in the following way.

Algorithm WSI. Given z^k, solve

Problem $(O_z k)$. Minimize the function

$$\tilde{F}(x,z^k) = F(x^k) + (x-x^k)^T F_x(x^k) + \frac{1}{2}(x-x^k)^T L_{xx}(z^k)(x-x^k)$$

subject to

$$\tilde{f}(x,y,z^k) = f(x^k,y) + (x-x^k)^T f_x(x^k,y) \leq 0, \quad y \in Y.$$

L(z) is the Lagrangian

$$L(z) = F(x) + \sum_{i=1}^{r} u^i f(x,y^i)$$

Making use of Theorem 7 and 8 the lemma below is easily proved.

Lemma. If the functions defining (O) are three times continuously differentiable then (5.3) holds with $\lambda = 2$ and the sequence $\{z^k\}$ converges quadratically.

We remark that the differentiability assumptions of this lemma can be weakened by replacing it by proper Lipschitz-conditions on the second partial derivatives of f.

References.

[1] L. Cromme: Eine Klasse von Verfahren zur Ermittlung bester nichtlinearer Tschebyscheff Approximationen, Numer. Math., 25 (1976), 447-459.

[2] A.V. Fiacco: Sensitivity analysis for nonlinear programming using penalty methods, Math. Programming, 10 (1976), 287-311.

[3] A.V. Fiacco and G.P. McCormick: Nonlinear programming: Sequential unconstrained minimization techniques, Wiley, New York, 1968.

[4] S.A. Gustafson and K.O. Kortanek: Numerical treatment of a class of semi-infinite programming problems, Nav. Res. Log. Quart., 20 (1973), 477-504.

[5] R. Hettich: A Newton method for nonlinear Chebyshev approximation, In: Approximation Theory, Lect. Notes in Math, 556 (1976), R. Schaback, K. Scherer, eds., Springer, Berlin-Heidelberg-New York, 222-236.

[6] R. Hettich: A comparison of some numerical methods for semi-infinite programmming, this volume.

[7] R. Hettich and H. Th. Jongen: Semi-infinite programming: conditions of optimality and applications, In: Optimization Techniques, Part 2, Lecture Notes in Contr. and Inform. Sciences, 7 (1978), J. Stoer, ed., Springer, Berlin-Heidelberg-New York, 1-11.

[8] R. Hettich and H.Th. Jongen: On first and second order conditions for local optima for optimization problems in finite dimensions, In: Methods of Operations Res. XXIII, R. Henn et al., Verlag A. Hain, Meisenheim a. Glan, 1977, 82-97.

[9] W. van Honstede: An approximation method for semi-infinite problems, this volume.

[10] I.M. Orthega and W.C. Rheinholdt: Iterative solution of nonlinear equations in several variables, Academic Press, New York, 1970.

[11] S.M. Robinson: Perturbed Kuhn-Tucker points and rates of convergence for a class of nonlinear programming algorithms, Math. Programming, 7 (1974), 1-16.

[12] R.B. Wilson: A simplicial algorithm for concave programming, Graduate School of Business Administration, Harvard University, Cambridge, Mass., 1963.

A comparison of some numerical methods for semi-infinite programming

R. Hettich
Institut für Angewandte Mathematik
Universität Bonn
Wegelerstr. 6
53 Bonn, West Germany

Abstract. Three ways to find approximate solutions of semi-infinite programming problems are considered: by discretization, exchange algorithms, and methods we call continuous because they make no use of discretization in the usual sense. Examples as well as theoretical considerations indicate that in general the first two ways only are useful in getting low accuracy approximations to a solution which preferably should be improved by a method of the third type.

1. Introduction.

Let $Y \subset \mathbb{R}^m$ be a compact, $X_0 \subset \mathbb{R}^n$ an open set and $F: X_0 \longrightarrow \mathbb{R}$, $f: X_0 \times Y \longrightarrow \mathbb{R}$ be given functions with appropriate differentiability properties. By X we denote the feasible region given by

$$X = \{x \in X_0 \mid f(x,y) \leq 0 \quad \text{for every} \quad y \in Y\} .$$

Problem (0). Minimize $F(x)$ subject to $x \in X$.

A solution $\hat{x}$ is called optimal or optimal solution and strictly optimal if $F(x) = F(\hat{x})$ implies $x = \hat{x}$.

Remarks. (1) In general, there may be locally optimal solutions as well, defined in the usual way. In the following, it does not matter if a locally or globally optimal solution is to be computed. Therefore, no such distinction is made.

(2) If $Y = \{y^1, \ldots, y^k\}$, then we have a problem with a finite number of constraints. Being mainly interested in the semi-infinite case, in the sequel Y

is assumed to be some continuum.

(3) For simplicity we consider only one function $f(x,y)$ in defining X (see however Problem (A) below). Without difficulties a finite number of functions $f^i : X_0 \times Y_i \longrightarrow \mathbb{R}$ could be considered and a finite number of equality constraints be added.

Most of the examples given below are (Chebyshev) approximation problems:

Problem (A). Let $Y \subset \mathbb{R}^m$ be a compact, $P \subset \mathbb{R}^n$ an open set, $\varphi : Y \longrightarrow \mathbb{R}$, and $a : P \times Y \longrightarrow \mathbb{R}$ be given functions. Find $\hat{x} = \begin{pmatrix} \hat{p} \\ \hat{d} \end{pmatrix} \in P \times \mathbb{R}$ such that

$$F(x) := d$$

is minimized subject to $x \in X = \{x \mid f^1(x,y) \le 0,\ f^2(x,y) \le 0,\ y \in Y\}$ with

$$f^i(x,y) := (-1)^i (\varphi(y) - a(p,y)) - d\ , \qquad i = 1,\ 2.$$

In the sequel we assume that an effective algorithm for solving problems with finitely many constraints is available. Then the following ways of computing approximate solutions of (0) will be considered:

Discretization. Problem (0) is solved with $\bar{Y} \subset Y$ instead of Y, $\bar{Y}$ a finite set.

Semi-continuous methods. (0) is solved for a sequence of finite subsets $Y_i \subset Y$ instead of Y. Y_{i+1} is some subset of $Y_i \cup \{y^{i1}, \ldots, y^{ir}\}$, y^{ij} local maxima of $f(x^i, y)$ on Y, x^i the solution on Y_i.

Continuous methods. Again the idea is to solve finite problems but now with constraints $f(x,\ y^i(x)) \le 0$, $y^i(x) \in Y$, rather than $f(x,\ y^i) \le 0$.

2. Discretization.

Let $\bar{Y} \subset Y$ be a finite set, $\bar{d} = \max_{y \in Y} \{\min_{\bar{y} \in \bar{B}} \|y-\bar{y}\|_2\}$. Then it is likely that for $\bar{d}$ sufficiently small, the solution on $\bar{Y}$ instead of Y will be an approximation to the one on Y sufficiently good for practice.

This point - crucial for the pratical relevance of semi-infinite programming - is considered in [13] by means of the following example:

Example 1. Approximate $\varphi(y) = \sin y_1 \sin y_2$ on $[0,1] \times [0,1]$ by rationals of the form

$$a(p,y) = \sum_{i=0}^{k} p_i(y_1^i + y_2^i) / \sum_{i=0}^{k} p_{k+1+i}(y_1^i + y_2^i).$$

Let $\bar{p}$ be the best approximation on $\bar{Y}$ with maximal deviation $\bar{d}$ and d on $\bar{Y}$ and Y resp. Naturally, $d \geq \bar{d}$. Let

$$\eta = (d - \bar{d})/\bar{d} .$$

Then, for instance, $\eta = 10^{-1}$ means that the approximation error can be improved with at most ten per cent

$\bar{Y}$ was chosen the set of points $\begin{pmatrix} y_{1i} \\ y_{2j} \end{pmatrix}$, $i,j = 1,\ldots,m$, y_{1i} (y_{2i} resp.), $i = 1,\ldots,m$, being the extremals of the Chebyshev polynomial T_{m-1} on $[0,1]$. Thus, $|\bar{Y}| = m^2$.

The problem on $\bar{Y}$ was solved by an improved version of a method described in [4], about two to three times faster than the original one. The algorithm works iteratively requiring the solution of a linear program with $2k+2$ variables and $2m^2$ constraints in each step. For this, a revised simplex procedure provided by Wetterling [17] was used.

For $k = 1, 3, 5$ (i.e. $n = 3, 7, 11$ in (A)), the results are as follows (computing time in seconds on an IBM 370/168):

$k = 1$:

m	10	20	30	40	50	60
time	2	7	15	30	50	70
η	10^{-1}	$3\cdot 10^{-2}$	10^{-2}	$6\cdot 10^{-3}$	$4\cdot 10^{-3}$	$3\cdot 10^{-3}$

k = 3 :

m	10	20	30	40	50	60
time	5	20	50	75	120	240
η	$2\cdot 10^{-1}$	$7\cdot 10^{-2}$	$3\cdot 10^{-2}$	$2\cdot 10^{-2}$	10^{-2}	$6\cdot 10^{-3}$

k = 5 :

m	10	20	30	40	50	60
time	13	50	130	300	500	800
η	$7\cdot 10^{-1}$	$2\cdot 10^{-1}$	$6\cdot 10^{-2}$	4.10^{-2}	$3\cdot 10^{-2}$	$2\cdot 10^{-2}$

Thus, increasing n and m entails an enormous increase in computing time but only a poor decrease in η.

We remark that for $n = 3$, $m = 5$ the solution obtained within 1 second was improved to $\eta = 10^{-12}$ within one second by Newton's method (Section 4). This indicates that it is much more advantageous to treat the continuous problem even if discretization would be sufficiently accurate in practice.

3. Semi-continuous methods.

Now we will consider methods based on the following idea:

In the i-th step a finite subset $Y_i \subset Y$ is given. Let x^i be a solution on Y_i. Determine points $y^1, \dots, y^r \in Y$ for which $f(x^i, y)$ assumes local maxima. Choose $Y_{i+1} \subseteq Y_i \cup \{y^1, \dots, y^r\}$ according to some given rules.

Methods of this type are very popular in linear Chebyshev approximation. For problems meeting Haar's condition the two algorithms of Remes are well-known. Here, always $|Y_i| = n+1$. In the first algorithm, which may be considered as a generalization of the Simplex algorithm to the semi-infinite case, one point is exchanged in each step and in the second one, being asymptotically equivalent to Newton's method [16], all points are exchanged in general.

Algorithms generalizing the first algorithm of Remes or, for semi-infinite programming problems, the Simplex algorithm are given for instance in [3], [8], [9], [11], [12], [14].

Watson [15] proposed an algorithm similar to the second one of Remes but now letting $Y_{i+1} = Y_i \cup \{y^1, \dots, y^r\}$ instead of exchanging points.

We will discuss the methods by means of an example for which all of them become essentially equivalent.

Example 2. (cf. [10]). The best approximation for $\varphi(y) = y^2$ in $[0,2]$ by $a(p,y) = p_1 y + p_2 e^y$ is given by $\hat{p} = (0.184.., 0.418..)^T$ with maximal error $0.538..$ in only the two points $\hat{y}^1 = 0.406..$, $\hat{y}^2 = 2$.

Consequently, the discrete problem on $\hat{Y} = \{\hat{y}^1, \hat{y}^2\}$ has a one-dimensional manifold $\{p = \hat{p} + \lambda q \mid q = (1, -0.27..)^T, \lambda \in \mathbb{R}\}$ of solutions. Note that there is no finite subset $\bar{Y}$ of $[0,2]$ such that the problem on $\bar{Y}$ has $\hat{p}$ as unique solution. In fact, solving the problem on $\bar{Y}$ with the Simplex method will always lead to a solution different from $\hat{p}$ because the solution found will be a vertex with at least three constraints active.

Starting with $Y_0 = \{0, 1, 2\}$, the algorithms proceed as follows:

error function $\varphi(y) - a(p^0, y)$.

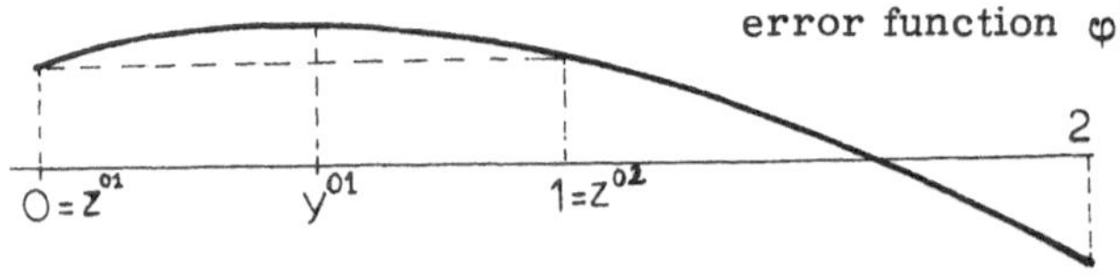

$Y_1 = \{0, 1, 2, y^{01}\}$

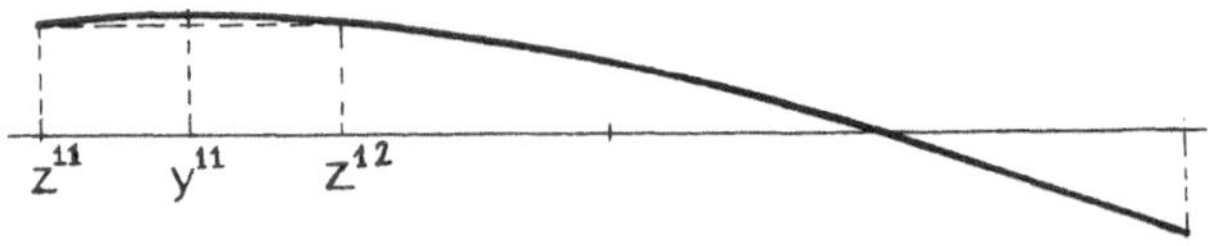

$Y_2 = \{0, 1, 2, y^{01}, y^{11}\}$

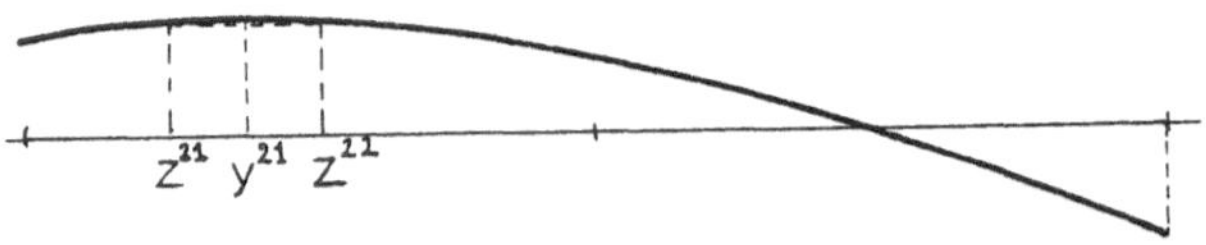

$Y_3 = \{0, 1, 2, y^{01}, y^{11}, y^{21}\}$.

Denoting by $\{z^{i1}, z^{i2}, z^{i3} = 2\}$ the points in the optimal reference, one observes that $\lim_{i\to\infty} z^{i1} = \hat{y}^1$ and $\lim_{i\to\infty} z^{i2} = \hat{y}^1$ as well. This implies that the problems become increasingly ill- conditioned.

Moreover increasing i, the new point $y^{i,1}$ will be about $\frac{1}{2}(z^{i,2} + z^{i,1})$. Thus, $\hat{y}^1$ is approximated by some sort of bisection, implying rather poor convergence.

Numerical examples confirming these unfavorable effects may be found in [1] and [2].

We emphasize that these disadvantages are inevitable at least if the maximal error of the best approximation is assumed in less than n+1 points on Y. As this happens to occur quite often if $Y \subset \mathbb{R}^m$, $m \geq 2$, this case may not be considered degenerate. For problem (0) the same applies if there are less than n points $\hat{y}^i \in Y$ such that $f(\hat{x}, \hat{y}^i) = 0$, $\hat{x}$ the solution in question.

4. Continuous methods (Newton - methods).

Finally we consider the Newton-methods described in [6], Sections 2 and 3. These - and the other methods in [6] - we call continuous because, contrary to those of Section 3, the approximations to $\hat{y}^i$ and $\hat{x}$ are not adapted independently.

As in [6] we assume the compact set $Y \subset \mathbb{R}^m$ to be given by

$$Y = \{y \mid g^j(y) \leq 0, \ j \in J\}, \tag{4.1}$$

with twice continuously differentiable functions $g^j : \mathbb{R}^m \longrightarrow \mathbb{R}$.

$\hat{Y}$ denotes the set of points being active for $\hat{x}$, the solution of (0), i.e.

$$\hat{Y} = \{y \mid f(\hat{x}, y) = 0\} \tag{4.2}$$

and we assume this set to be finite:

$$\hat{Y} = \{\hat{y}^1, \ldots, \hat{y}^r\}. \tag{4.3}$$

For each $y \in Y$ a set $J(y)$ of indices is given by

$$J(y) = \{j \in J \mid g^j(y) = 0\}. \tag{4.4}$$

In the sequel we repeatedly refer to definitions, theorems, and relations in [6]. First, recall the method of [6, Section 2], were $\hat{x}, \hat{y}^i, \hat{u}^i, \hat{w}^{ij}$, $j \in J(\hat{y}^i)$, $i=1,\dots,r$, are to be determined from the nonlinear system [6, (2.13) - (2.16)]

$$F_x(x) + \sum_{i=1}^{r} u^i f_x(x, y^i) = 0 \qquad (4.5)$$

$$f(x, y^i) = 0, \quad i = 1,\dots,r \qquad (4.6)$$

$$f_y(x, y^i) - \sum_{j \in J(\hat{y}^i)} w^{ij} g_y^j(y^i) = 0, \quad i = 1,\dots,r \qquad (4.7)$$

$$g^j(y^i) = 0, \quad j \in J(\hat{y}^i),\ i=1,\dots,r. \qquad (4.8)$$

Note that in order to formulate (4.5) - (4.8) we must have the following information:

Assumption on starting point. The number r of active points $\hat{y}^i$ and the sets $J(\hat{y}^i)$ must be given.

For instance, if $Y = [0,1] \times [0,1] \subset \mathbb{R}^2$, besides r we must know how many of the $\hat{y}^i$ are in the interior of Y, how many are situated on each of the four faces, and which of the vertices belong to $\hat{Y}$.
Furthermore, to be sure of superlinear convergence, due to [6, Thm. 5] we should have:

Assumption on problem. Property (B) [6, Section 2] is given.

We will discuss, how restrictive the above assumptions prove to be in practice. Let us start with the assumption on the starting point. Here, numerical experience seems to be the only way to get information about this point. Tests have been made with several types of approximation problems indicating that discretization with only a few points as well as performing a few steps of some semi-continuous method rather quickly leads to an approximation sufficient for getting the required information. We give some examples:

Example 3. Approximate $f(x) = \sqrt{x}$ on $[0.25, 1]$ by the H-polynomial (cf.[5]) $a(p,x) = \sigma((p_0 x^2 + p_1 x + p_2)^2 + p_3$, $\sigma \in \{-1, 1\}$.

The solution of the discretized problem on the five points $x^i = 0.25 + i\,0.75/4$,

$i = 0, \dots, 4$, leads to an error function with four local extrema. Taking these as starting values for ϑ^i, $i = 1, \dots, 4$(i.e. $r = 4$) is sufficient for Newton's method to converge. After three steps an approximation is obtained with relative accuracy $\eta < 10^{-5}$ for the error.

Other examples for approximation on $Y = [a, b] \subset \mathbb{R}$ gave similar results. Now, let us consider an example where $Y \subset \mathbb{R}^2$.

Example 4. In [13] it is shown that for Example 1 and $k = 3$(i.e. $n = 7$) already the discrete approximation on $\bar{B}$ for $m = 5$(i.e. 25 discretization points) yields the required information about r, $J(\vartheta^i)$. Starting with the discrete approximation for $m = 5$, Newton's method gave an approximation with relative accuracy $\eta < 10^{-12}$ for the error within 5 iterations and a computing time of about 1 sec.

We remark that this good result is obtained nevertheless the matrix of the linearized system (4.5) - (4.8) is singular in the solution point. We note further that in this example $r = 6 < 8 = n+1$ (due to symmetry only half of the square was considered).

Further examples with linear problems and semi-continuous methods for getting a starting point and showing a similarly good behaviour may be found in [1] and [2].

Alltogether, numerical experience indicates that, if discrete problems entailing a low number of discretization points can be solved effectively, it is rather easy in practice to find a good starting point for Newton's method. Especially this is true for general linear and rational approximation problems (A).

Finally we remark that methods from [6] based on (FO') do not require the assumptions on the starting point (cf. [7]). For general nonlinear problems further research on this point would be desirable.

Now let us consider Property (B), sufficient for local superlinear convergence of Newton's method applied to (4.5) - (4.8) and the Newton method based on problem (FO) as well (cf. [6, Section 3]). As the latter is more suitable for the following discussion, we recall it briefly.

Problem (FO). Minimize $F(x)$ subject to $f(x, y^i(x))$, $i = 1,\dots,r$, $x \in \hat{X}$, $\hat{X}$ some neighborhood of $\hat{x}$.

Here, $y^i(x)$ are continuously differentiable functions $\hat{X} \longrightarrow Y$, such that $y^i(\hat{x}) = \hat{y}^i$ and such that in some neighborhood $y^i(x)$ is the only local maximum of $f(x,y)$. Especially, (4.7), (4.8) hold for $y^i = y^i(x)$ for each $x \in \hat{X}$.

Then the method was:

Step 1. Given x^0, $u^{0,i}$, $y^{0,i} = y^i(x^0)$. Perform one step of Newton's method to solve

$$F_x(x) + \sum_{i=1}^{r} u^i f_x(x, y^i(x)) = 0 \qquad (4.9)$$

$$f(x, y^i(x)) = 0 \quad , \quad i = 1,\dots,r \qquad (4.10)$$

yielding new approximations x^1, $u^{1,i}$. Let $x^0 = x^1$, $u^{0,i} = u^{1,i}$.

Step 2. Compute $y^i(x^0)$, $i = 1,\dots,r$ and continue with Step 1.

An advantage of this method above Newton's method applied to (4.5) - (4.8) is that the determination of the maxima of $f(x,y)$ on Y is performed independently. Note, however, that the change of y^i dependent on x enters (4.9),(4.10).

Remark. If $r = n$, then (4.10) is a system of n equations for $x \in \mathbb{R}^n$, i.e. (4.9) is unnecessary. It is easily shown that in this case the multiple exchange algorithm of Watson (cf. Section 3) is essentially equivalent to the above method. For linear approximation problems (A), if Haar's condition holds, this has been observed first by Wetterling [16], who proved the equivalence of the second algorithm of Remes to Newton's method. As the above method - contrary to the exchange method - continues to be superlinearly convergent for $r < n$ (Property (B) given) it may be considered the appropriate generalization of the second algorithm of Remes.

We will study two implications of Property (B):

(1) If Property (B) holds, then $\hat{x}$ is a locally unique optimal point.

(2) For $i = 1,\dots,r$, the matrix

$$\hat{M}^i := f_{yy}(\hat{x},\hat{y}^i) - \sum_{j\in J(\hat{y}^i)} \hat{w}^{ij} g^j_{yy}(\hat{y}^i) \tag{4.11}$$

is negative definite on the subspace $T_i = \{\xi \mid \xi^T g^j_y(\hat{y}^i) = 0,\ i = 1,\dots,r\}$.

(1) seems to indicate, that if $\hat{x}$ is not strictly optimal, the methods will fail to converge. The following consideration and example show that this is not at all true in general. From [6, Thm. 1] we know that if $\hat{x}$ is strictly optimal for (0), then it is for (FO). The following example shows that the converse is not generally true:

Example 5. Approximate $f(x) \equiv 1$ on $[0,\ 0.75]$ by $a(p,x) = 8px - 8x^2$, $p \in \mathbb{R}$. It is easily seen that every $p \in [0.75,\ 1]$ yields a best approximation. Now, solving the discretized problem with the Simplex-method leads to an approximation with maximal (discrete) error in at least two points.

For instance, solve the problem on $\bar{Y} = \{0,\ 0.4\}$. This gives $p^0 = 1.025$, $d^0 = 1$, $u^{0,1} = 1$, $u^{0,2} = 0$, $y^{0,1} = 0.5125$, $y^{0,2} = 0$. It follows $y^1(p) = 0.5p$, $y^2(p) \equiv 0$. Therefore, (FO) becomes:
Minimize d subject to $1-d = 0$, $1-2p^2+d = 0$. Clearly, $\hat{p} = 1$ is a locally unique solution to (FO), and Newton's method is easily seen to converge quadratically to this solution.

Remark. The above consideration indicates that Property (B) is a condition for convergence unnecessarily strong. Indeed it is straightforward to prove convergence if Problem (FO) has a property analoguous to (B) for finite problems.

Now let us consider the case that during the iteration we arrive at a point $\tilde{x}$, $\tilde{y}^i$, $\tilde{u}^i$, $\tilde{w}^{ij}$, where, for some $i \in \{1,\dots,r\}$ the matrix $\tilde{M}^i$, defined analogously to (4.11), is singular or nearly singular. For simplicity we assume $J(\hat{y}^i) = \emptyset$, i.e. $\tilde{y}^i$ is an interior point of Y. Further, we assume that an arc $y^i(x)$, continuously differentiable, exists such that $y^i(\tilde{x}) = \tilde{y}^i$ and such that $y^i(x)$ is a local maximum of $f(x,y)$ with respect to Y, not necessarily locally unique.

Application of Newton's method to (4.9), (4.10) with starting point $\tilde{x}, \tilde{u}^i$ requires the solution of a linear system with matrix of coefficients

$$\begin{pmatrix} A & G \\ G^T & O \end{pmatrix}$$

where

$$A = F_{xx}(\tilde{x}) + \sum_{i=1}^{r} \tilde{u}^i [f_{xx}(\tilde{x}, \tilde{y}^i) + f_{xy}(\tilde{x}, \tilde{y}^i) y_x^i(\tilde{x})] \tag{4.12}$$

and

$$G = (f_x(\tilde{x}, y^1(\tilde{x})), \dots, f_x(\tilde{x}, y^r(\tilde{x}))).$$

Thus, computation of $f_{xy} y_x^i$ in (4.12) is required (here and in the sequel the arguments $\tilde{x}, \tilde{y}^i$ are omitted). From $f_y(x, y^i(x)) \equiv 0$ it follows that

$$\frac{d}{dx} f_y(x, y^i(x)) = f_{yx}(x, y^i(x)) + f_{yy}(x, y^i(x))\, y_x^i(x) \equiv 0 \tag{4.13}$$

If f_{yy} is regular, this implies $y_x^i = -f_{yy}^{-1} f_{yx}$ and $f_{xy} y_x^i = -f_{xy} f_{yy}^{-1} f_{yx}$. Therefore, for f_{yy} singular or nearly singular (4.12) is undefined.

Now let $\lambda_1, \dots, \lambda_m$ be the eigenvalues of f_{yy} and $v_1, \dots, v_m$ an orthonormal set of eigenvectors. Then, with real valued functions t_j we may write

$$y^i(x) = \sum_{j=1}^{m} v_j\, t_j(x) + \hat{y}^i \tag{4.14}$$

implying

$$y_x^i = \sum_{j=1}^{m} v_j\, t_{jx}^T . \tag{4.15}$$

From (4.13) we obtain

$$f_{yx} = -f_{yy} \sum_{j=1}^{m} v_j\, t_{jx}^T = -\sum_{j=1}^{m} \lambda_j\, v_j\, t_{jx}^T$$

and

$$f_{xy}\, y^i_x = -\sum_{j=1}^{m} \lambda_j\, t_{jx}\, t^T_{jx} . \tag{4.16}$$

(4.16) shows that $f_{xy}\, y^i_x$ in (4.12) is independent of $t_j(x)$ if $\lambda_j = 0$. Therefore, in computing (4.12) it does not mean if we take

$$\bar{y}^i(x) = \hat{y}^i + \sum_{j=1}^{m'} v_j\, t_j(x) \tag{4.17}$$

instead of $y^i(x)$ (4.14), where it is assumed that $\lambda_j = 0$ exactly for $j = m'+1, \dots, m$. From (4.13) we get for $j = 1, \dots, m'$

$$0 = v_j^T f_{yx} + v_j^T f_{yy}\, y^i_x = v_j^T f_{yx} + \lambda_j v_j^T y^i_x = v_j^T f_{yx} + \lambda_j t^T_{jx}$$

thus

$$t_{jx} = -\frac{1}{\lambda_j} f_{xy}\, v_j .$$

Therefore, by (4.16), we have

$$f_{xy}\, y^i_x = -\sum_{j=1}^{m'} \frac{1}{\lambda_j} f_{xy}\, v_j\, v_j^T f_{yx} . \tag{4.18}$$

The formula (4.18) can be applied if f_{yy} is singular or not. In most practical problems m is a small number (mostly $m < 3$) so that the eigenvalue analysis can be performed very quickly.

Example 6. Approximate $\varphi(y) = y_1^2 + y_2^2$ in $[0,\sqrt{2}] \times [0,\sqrt{2}]$ by $a(p,y) = p_1\sqrt{y_1^2+y_2^2} + p_2 \exp(\sqrt{y_1^2+y_2^2})$. (Letting $y = \sqrt{y_1^2+y_2^2}$ we obtain Example 2). The unique best approximation is $\hat{p} = \begin{pmatrix}0.148..\\0.418..\end{pmatrix}$ and, for $\hat{p}$, maximum error is attained in the points $\begin{pmatrix}\sqrt{2}\\\sqrt{2}\end{pmatrix}$ and $y \in \hat{S}$, $\hat{S} = \{y \mid \sqrt{y_1^2+y_2^2} = 0.406..\}$.

To solve the problem start by solving the discretized problem on the set $\bar{Y} = \{ \begin{pmatrix}0\\0\end{pmatrix}, \frac{1}{2}\begin{pmatrix}\sqrt{2}\\\sqrt{2}\end{pmatrix}, \begin{pmatrix}\sqrt{2}\\\sqrt{2}\end{pmatrix} \}$ yielding $p^0 = \begin{pmatrix}0.306..\\0.403..\end{pmatrix}$. Local extrema of the corresponding errorcurve occur in $y^1 = \begin{pmatrix}\sqrt{2}\\\sqrt{2}\end{pmatrix}$ and for $y \in S_0$, $S_0 = \{y \mid \sqrt{y_1^2+y_2^2} = 0.479..\}$.

Take for instance $y^2 \in S_0$, $y^2 = \begin{pmatrix}0.339..\\0.339..\end{pmatrix}$. Then $f_{yy}(x^0, y^2)$ has eigenvectors $\frac{1}{2}\binom{\sqrt{2}}{\sqrt{2}}$ and $\frac{1}{2}\binom{\sqrt{2}}{\sqrt{2}}$, the latter corresponding to an eigenvalue 0. Now, letting $y^1(x) \equiv \binom{\sqrt{2}}{\sqrt{2}}$, $y^2(x) = \frac{1}{2}\, t(x)\binom{\sqrt{2}}{\sqrt{2}}$, the method converges without any difficulty.

The above analysis shows that by certain additional strategies the method is applicable also if certain parts of Property (B) do not hold and continues to be superlinearly convergent. So the range of applicability of the method is much broader than may be expected in first instance.

References.

[1] D.O. Andreassen and G. A. Watson: Linear Chebyshev approximation without Chebyshev sets, BIT, 16 (1976), 349 - 362

[2] W. Braun: Ein Verfahren zur linearen Tschebyscheff-Approximation auf mehrdimensionalen Bereichen, Diplomarbeit Universität Bonn, 1978

[3] E.W. Cheney and A.A. Goldstein: Newton's method for convex programming and Tchebyscheff approximation, Numer. Math., 1 (1959), 253-268

[4] L. Collatz und W. Krabs: Approximationstheorie, Teubner-Verlag, Stuttgart, 1973

[5] R. Hettich: A Newton method for nonlinear Chebyshev approximation, In: Approximation Theory, Lect. Notes in Math., 556 (1976), R. Schaback and K. Scherer, eds., Springer, Berlin-Heidelberg-New York, 222 - 236

[6] R. Hettich and W. van Honstede: On quadratically convergent methods for semi-infinite programming, this volume

[7] W. van Honstede: An approximation method for semi-infinite problems, this volume

[8] K.H. Hoffmann and A. Klostermaier: A semi-infinite programming procedure, In: Approximation Theory II, 1976, G.G. Lorentz et al., eds., Academic Press, New York - San Francisco - London, 379-389

[9] P.J. Laurent et C. Carasso: Un algorithme pour la minimisation d'une fonctionelle convexe sur une variété affine, preprint Grenoble, 1973

[10] M.R. Osborne and G.A. Watson: A note on singular minimax approximation problems, J. Math. Anal. Appl., 25(1969), 692 - 700

[11] R. Schaback and D. Braess: Eine Lösungsmethode für die lineare Tschebyscheff-Approximation bei nicht erfüllter Haarscher Bedingung, Computing, 6 (1970), 289 - 294

[12] E. Schäfer: Ein Konstruktionsverfahren bei allgemeiner linearer Approximation, Numer. Math., 18 (1971), 113 - 126

[13] G. Speich: Ein Algorithmus zur Lösung diskreter allgemeiner rationaler Approximationsprobleme, Diplomarbeit Universität Bonn, 1978

[14] H.J. Töpfer: Tschebyscheff-Approximation bei nicht erfüllter Haarscher Bedingung, ISNM, 7 (1967), 71 - 89

[15] G.A. Watson: A multiple exchange algorithm for multivariate Chebyshev approximation, SIAM J. Numer. Anal., 12 (1975), 46 - 52

[16] W. Wetterling: Procedure RVSA (revidierte Simplexmethode A), preprint, Enschede, 1974

[17] W. Wetterling: Anwendung des Newtonschen Iterationsverfahrens bei der Tschebyscheff-Approximation, insbesondere mit nichtlinear auftretenden Parametern, MTW, 10 (1963), Teil I: 61 - 63, Teil II: 112 - 115

An approximation method for semi-infinite problems

W. van Honstede
Institut für Angewandte Mathematik
Universität Bonn
Wegelerstr. 6
5300 Bonn, West Germany

Abstract. Numerical methods for general non-linear semi-infinite problems, which make use of linear approximations, need strong assumptions to ensure local quadratic convergence.
An algorithm, solving a sequence of linearly constrained semi-infinite problems, is proposed which under weaker assumptions converges locally quadratic too.

1. Introduction. Consider the following finite programming problem:

Problem (FP). Given an open set $X_o \subset \mathbb{R}^n$, a finite set I of indices and three times continuously differentiable functions $F: X_o \longrightarrow \mathbb{R}$, $f^i: X_o \longrightarrow \mathbb{R}, i \in I$.
Minimize $F(x)$ subject to $x \in X$ where the feasible region X is defined as:

$$X = \{ x \in X_o \mid f^i(x) \leq 0 , \quad i \in I = \{1, --, r\} \} ,$$

(FP) may be treated by making linear approximations to the given functions. That is generate a sequence $\{x^k\}$, x^{k+1} being the solution of the linear programming problem:

Problem (LP_k). Minimize the function

$$\bar{F}(x, x^k) = F(x^k) + (x-x^k)^T F_x(x^k)$$

subject to

$$\bar{f}^i(x, x^k) = f^i(x^k) + (x-x^k)^T f^i_x(x^k) \leq 0 , \quad i \in I.$$

(LP_k) can be solved using the Simplex method.
A sufficient condition for locally quadratic convergence to an optimal solution $\hat{x}$ of (FP) includes the assumption that $\hat{x}$ is locally strongly unique (For Chebyshev-approximation this is proved by Cromme [1]).

Definition 1. $\hat{x}$ is a locally strongly unique solution of (FP) if for any $x \in X \cap U(\hat{x})$ ($U(\hat{x})$ some neighbourhood of $\hat{x}$) $F(x) - F(\hat{x}) \geq \alpha \| x - \hat{x} \|$, $\alpha > 0$ independent of x, $\| \cdot \|$ some vectornorm.

If this rather restrictive assumption is not satisfied, the following simple example shows the fundamental difficulty inherent to the approach above.

Minimize $F(x) = +(x_1 - 1/2)^2 + (x_2 + 1/2)^2$ on the square $0 \leq x_1, x_2 \leq 1$.

The optimal solution is $\hat{x} = (1/2, 0)^T$. Then, replacing $F(x)$ by some linear function, the Simplex method applied to this linear problem will always terminate in a vertex. So, without additional strategies like stepreduction, we will never reach $\hat{x}$.

Therefore other possibilities have been considered, for instance by Wilson ([5]) who solves (FP) by means of a sequence of quadratic programming problems:

Given (x^k, u^k), compute the optimal solution x^{k+1} of

Problem (FP_k'). Minimize the function

$$\overset{*}{F}(x, x^k, u^k) = F(x^k) + (x-x^k)^T F_x(x^k) + 1/2(x-x^k)^T L_{xx}(x^k, u^k)(x-x^k)$$

subject to

$$\overset{*}{f}{}^i(x, x^k) = f^i(x^k) + (x-x^k)^T f_x^i(x^k) \leq 0, \quad i \in I$$

where L is the Lagrangian

$$L(x, u) = F(x) + \sum_{i \in I} u^i f^i(x),$$

using a minimization method which calculates x^{k+1}, u^{k+1} solving

$$L_x(x, u) = \overset{*}{F}_x(x, x^k, u^k) + \sum_{i \in I} u^i \overset{*}{f}{}_x^i(x, x^k) = 0.$$

Then, in some neighbourhood of $(\hat{x}, \hat{u})$, the sequence (x^k, u^k) converges quadratically if a sufficient second-order optimality condition holds.

In the following section we will generalize this method to the semi-infinite problem:

Problem (O). Let $X_o \subset \mathbb{R}^n$ be an open, $Y \subset \mathbb{R}^m$ a compact set and $F: X_o \longrightarrow \mathbb{R}$, $f: X_o \times Y \longrightarrow \mathbb{R}$, $g^j: \mathbb{R}^m \longrightarrow \mathbb{R}$, $j \in J$, $|J| < \infty$, be three times continuously differentiable functions.

Minimize $F(x)$ subject to $x \in X$, where the feasible region X is given by

$$X = \{ x \in X_o \mid f(x,y) \leq 0 , y \in Y \}$$

and

$$Y = \{ y \mid g^j(y) \leq 0, j \in J \}.$$

2. Formulation of the method.

From now on Property (B) of Section 2, [4], is assumed to hold. That is, some second-order sufficient condition for $\hat{x}$ to be strictly locally optimal for (O) is given.

The following algorithm, a generalization of Wilson's method to semi-infinite problems, is also formulated in Section 6 of [4] as an example of the general algorithm (AG) described in Section 5 of that paper.

Algorithm (WSI).

Start: Let $z^k = (x^k, u^{ik}, y^{ik})$, $i = 1, --, r$, $k = 0$, be given.
Step (k): For $k \geq 0$, compute the optimal solution z^{k+1} of

Problem $(O_z k)$. Minimize the function

$$\tilde{F}(x, z^k) = F(x^k) + (x-x^k)^T F_x(x^k) + 1/2 \ (x-x^k)^T L_{xx}(z^k) \ (x-x^k)$$

subject to

$$\tilde{f}(x, y, z^k) = f(x^k, y) + (x-x^k)^T f_x(x^k, y) \leq 0 , \ y \in Y$$

where L is the Lagrangian

$$L(z) = F(x) + \sum_{i=1}^{r} u^i f(x, y^i) .$$

I.e. problem (O) is treated by means of a sequence of semi-infinite problems $(O_z k)$ of a simpler structure than (O).
In [4] the following theorem is proved.

Theorem 1. If Property (B) holds, then
(i) for each k, z^{k+1} satisfies Property (B) for $(O_z k)$.

(ii) the sequence $\{z^k\}$ converges quadratically.

To solve the semi-infinite problem $(O_z k)$, we consider the finite optimization problem

Problem (FO'_k). Minimize

$$\tilde{F}(x, z^k) = F(x^k) + (x-x^k)^T F_x(x^k) + 1/2\,(x-x^k)^T L_{xx}(z^k)(x-x^k)$$

subject to $x \in \tilde{X}$ ($\tilde{X}$ some neighbourhood of x^{k+1}) and

$$\tilde{f}(x, y^i(x), z^k) = f(x^k, y^i(x)) + (x-x^k)^T f_x(x^k, y^i(x)) \leq 0$$

$$\tilde{f}(x, y^l, z^k) = f(x^k, y^l) + (x-x^k)^T f_x(x^k, y^l) \leq 0, \quad y^l \in \tilde{Y}.$$

$\tilde{Y}$ a finite subset of Y and L the Lagrangian above.

Then from Theorem 1 (i) above and Theorem 1 of [4] we have:

Theorem 2. x^{k+1} is strictly locally optimal for $(O_z k)$ if and only if x^{k+1} is strictly locally optimal for (FO'_k).

This result enables us to compute the optimal solution z^{k+1} of $(O_z k)$ for instance in the following way.

Step k. Start with $x^s = x^k$, $u^{is} = u^{ik}$, $i=1,--r$, $s = 0$ (For $k = 0$ x^s, u^{is} can be chosen arbitrarily).

(i) Maximize the function $\tilde{f}(x^s, y, z^k)$, $y \in Y$, giving y^{is}, $i=1,--r$.

(ii) Test for convergence. Go to step (k+1) with $z^{k+1} = z^s$ or set $s = s+1$.

(iii) To compute x^s, u^{is}, solve the finite optimization problem

Problem $(FO'_k)^{s-1}$. Minimize

$$\bar{F}(x, z^{s-1}, z^k) = \tilde{F}(x^{s-1}, z^k) + (x-x^{s-1})^T \tilde{F}_x(x^{s-1}, z^k) + 1/2(x-x^{s-1})^T \tilde{L}_{xx}(z^{s-1}, z^k)(x-x^{s-1})$$

subject to

$$\tilde{f}(x, y^{i,s-1}, z^k) = f(x^k, y^{i,s-1}) + (x-x^k)^T f_x(x^k, y^{i,s-1}) \leq 0, \; i = 1, --, r$$

$$\tilde{f}(x, y^l, z^k) = f(x^k, y^l) + (x-x^k)^T f_x(x^k, y^l) \leq 0, \quad y^l \in \tilde{Y}.$$

Here $\tilde{L}$ is the Lagrangian

$$\tilde{L}(z, z^k) = \tilde{F}(x, z^k) + \sum_{i=1}^{r} u^i \tilde{f}(x, y^i(x), z^k)$$

and $\tilde{L}_{xx}$ can be shown to be

$$\tilde{L}_{xx}(z, z^k) = \sum_{i=1}^{r} u^i f_{xy}(x^k, y^i) \, y^i_x(x) + L_{xx}(z^k)$$

For every i, the (m x n)- matrix y^i_x can be computed by solving a linear system of equations (cf. Section 3, [4]).

Go back to step (i).

If for every s, $(FO'_k)^{s-1}$ is solved such that

$$\bar{L}_x(z, z^{s-1}, z^k) = \bar{F}_x(x, z^{s-1}, z^k) + \sum_{i=1}^{r} u^i \tilde{f}_x(x, y^{i, s-1}, z^k) = 0$$

then, for z^k, z^s close enough to $\hat{z}$, step k(iii) is equivalent with one iteration of Newton's method applied to the nonlinear system of equations

$$\tilde{F}(x, z^k) + \sum_{i=1}^{r} u^i \tilde{f}_x(x, y^i(x), z^k) = 0$$

$$\tilde{f}(x, y^i(x), z^k) = 0 , \quad i=1, --, r$$

Therefore, locally step k is equivalent with the Newton method of Section 3, [4] , with respect to $(O_z k)$ instead of (O) and quadratic convergence of the sequence $\{z^s\}$ is immediate.

An advantage of this method above linear approximation is, apart from the quadratic convergence property, the simple way we are able to deal with fluctuations in the number of active constraints $\tilde{f}(x, y^i(x), z^k)$, $i=1, --, r$ $(r \leq n)$. In the next section, where some remarks on the practical use of algorithm (WSI) are made, a more detailed discussion of this point will be given.

3. Computational considerations.

Given z^k (k=0), x^s (s=0), in step k(i) Newton's method, with starting point obtained by maximizing $\tilde{f}(x^s, y, z^k)$ on a discrete set of points y, is applied to calculate the local maxima y^{is}, i=1, --, r. As Newton's method may diverge, another method (direct search method, gradient method) should be used, if convergence is not achieved in less than a given number p of iterations (p depends on the problem (O) in question).

For $s=0$, u^{is}, $i=1,--,r$, are set to 1 with r equal to the number of local maxima y^{is} met up to now. Then $(FO'_k)^{s-1}$ is solved making use of an optimizationmethod for linearly constrained problems. Naturally a numerically stable method like [2] is preferable.

Because, for $r<n$, $(FO'_k)^{s-1}$ may be ill-defined, we add at least $(n-r)$ additional fixed points y^l, making the number of constraints $\tilde{f}$ greater or equal to n.

The flexibility of the algorithm with respect to changes in the number of active constraints $\tilde{f}(x, y^i, z^k)$ (we call these points y^i "active" points) is based on the following strategies:

(i) If a constraint $\tilde{f}(x, y^l, z^k)$ becomes active, we treat y^l in the following steps as a "normal" y^i.

(ii) Because during the computations new local maxima y^i may arise, for each iteration s, in step k(i) a discretization of $\tilde{f}$ is performed. If the local maxima $\tilde{y}^i$, found by the discrete search, are not in a neighbourhood of already existing y^i , a few iterations of Newton's method with starting points $\tilde{y}^i$ will generate the new local maxima.

(iii) If local maxima y^i converge to each other they simply are joined.

For the testproblems of Section 4 in step k(iii) an active set strategy of Fletcher (a subroutine of the Harwell-library) and in step k(i) only Newton's method is used.

4. Numerical results.

The problems presented here are simplifications of a mathematical model for air-pollution control (cf. [3]).

Given a control region with plants P_i , this model is used to minimize the costs which occur if the production of the plants P_i must be reduced to reach compliance with given air-quality standards for the ground level Y. In this section we consider a control region with three plants, all plants emitting the same pollutant. Further let $c_i(y)$, $i=1, 2, 3$, be the contribution of P_i to the concentration of the pollutant in point y of Y, x_i $(i=1, 2, 3)$ be the production

reduction factor with respect to P_i $(0 \leq x_i \leq 1)$ and $F(x)$ be the associated cost function.
The maximum permitted concentration in Y is fixed at 1/2.

In the first example we assume that the costs increase linearly with the re - duction rate x.

Example 1. Minimize the function

$$F(x) = 2x_1 + 4x_2 + x_3$$

subject to

$$f(x,y) = \sum_{i=1}^{3} (1-x_i)\, c_i(y) - 1/2 \leq 0, \quad y \in Y = ([-1,4],[-1,4])^T$$

Here $c_i(y)$, $i=1,2,3$, is given by

$$c_1(y) = (1/y_1) \exp((-1/y_1)(1 + (y_2-1)^2)) \quad \text{for} \quad y_1 > 0$$

$$c_2(y) = (1/y_1) \exp((-1/y_1)(2 + y_2^2/4)) \quad \text{for} \quad y_1 > 0$$

$$c_3(y) = (1/(y_1-2)) \exp((-1/(y_1-2))(1 + (y_2+1)^2)) \quad \text{for } y_1 > 2$$

$$c_{1,2,3}(y) = 0 \quad \text{elsewhere}$$

The positions of P_i are indicated in Fig. 1. For more details see [3].

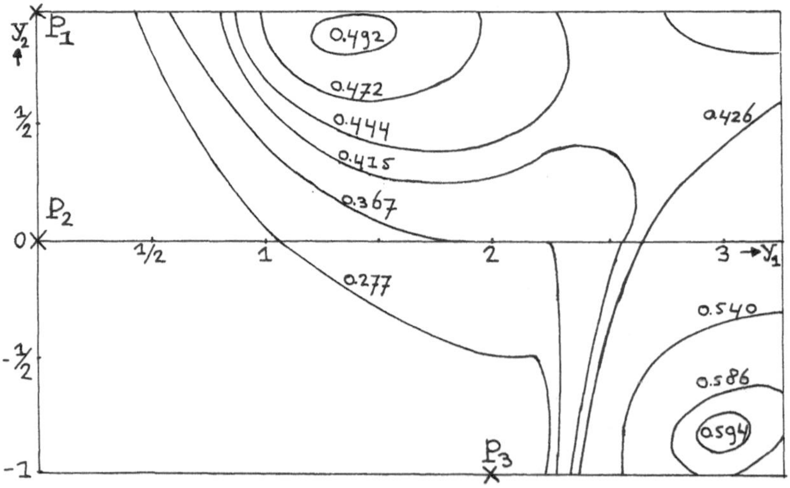

Fig 1. The contour-lines of $\sum_{i=1}^{3} c_i(y)$

Because the functions F and f of Problem 1 are linear with respect to x, z^k (k=0) may be chosen arbitrarily and algorithm (WSI) will terminate after one iteration (k).

a) Starting point: $x^s = (1., 0., 1.)^T$, s=0.

Convergence - condition: relative accuracy 10^{-3} for z

	s=0	s=1	s=2	s=5
x^s	1. 0. 1.	0. 0. 0.	0. 0. 0.27176	0. 0. 0.27527
$\bar{F}(x, z^{s-1}, z^k)$	—	0.36788	0.27490	0.27527
$y^{1,s}$	2. 0.	4. -1.	4. -1.	4. -1.
$y^{2,s}$	— —	3.01520 -0.83370	3.03280 -0.75571	3.03477 -0.75369
$y^{3,s}$	— —	1.35979 0.90613	1.35959 0.90604	1.35959 0.90604

Execution time: about 5 sec.
(IBM 370/168)

Explanation

Starting algorithm (WSI), in step k(i) $y^{1,s}$, s=0, is computed as described in Section 2. To make sure that $(FO_k')^{s-1}$, s=1, is well-defined, two points y^l ($y^1 = (1/4, 11/4)^T$, $y^2 = (11/4, 11/4)^T$) are added.
The optimal solution x^1 of $(FO_k')^{s-1}$ is an unconstrained minimum. I.e. no points y^i or y^l are active.
Returning to step k(i), Newton's method with starting point $y^{1,0}$ generates a vertex of the feasible region Y. By a discrete search with respect to $\tilde{f}(x^1, y, z^k)$ we find approximated new local maxima $\tilde{y}^{i1}$ ($\tilde{y}^{1,1} = (3., -0.8)^T$, $\tilde{y}^{2,1} = (1.4, 1.)^T$) and using Newton's method we obtain the points $y^{i,1}$, i=2, 3.
For the next iterations (s≥2) no changes occur in the number of points y^{is} and for each iteration s only point $y^{2,s}$ is active.
This example illustrates clearly how the number of points y^i can change during the computation and how the algorithm deals with it.

b) Starting point: $x^s = (0.1, 0.2, 0.5)^T$, s=0

Convergence - condition: relative accuracy 10^{-3} for x

	s=0	s=1	s=2	s=5
x^s	0.1 0.2 0.5	0. 0. 0.26946	0. 0. 0.27526	0. 0. 0.27527
$\bar{F}(x, z^{s-1}, z^k)$	—	0.25162	0.27521	0.27527

Execution time: about 3 sec.

A more realistic approach may be a cost function which is nonlinearly dependent of one or more x_i. Therefore the following example.

Example 2. Minimize

$$F(x) = 2x_1 + 4x_2 + x_3 + 30x_3^2 + 30x_3^3$$

subject to the constraints of example 1.

Starting point: $x^k = x^s = (0., 0.4, 0.1)^T$, $y^{1,k} = (3., -0.8)^T$, $u^{1,k} = 25.$, $k = s = 0$
Convergence - condition: relative accuracy 10^{-3} for z.

k/s:	0/0	0/1	0/2	0/3	1/1	1/2	2/1
x^s	0. 0.4 0.1	0. 0.35457 0.11224	0. 0.35496 0.11208	0. 0.35499 0.11207	0. 0.35531 0.11192	0. 0.35531 0.11192	0. 0.35533 0.11192
$\bar{F}(x, z^{s-1}, z^k)$	—	1.95092	1.95097	1.95097	1.95102	1.95097	1.95108

Execution time: about 7 sec.

The "optimal" solution z^2 contains one active point $y^{1,2} = (3.03630, -0.82092)^T$ with Lagrange-multiplier $u^{1,2} = 24.80679$. The last three iterations are necessary for the sequences $\{u^{is}\}$ to met the convergence-condition.

For example 3 we have changed the position of P_2 from (2., 0.) to (1., 0.) and the contribution of P_1 is doubled. The cost function is the one of example 1.

Example 3. Minimize

$$F(x) = 2x_1 + 4x_2 + x_3$$

subject to $0 \leq x^i \leq 1$, i=1,2,3, and

$$f(x,y) = \sum_{i=1}^{3} (1-x_i)\, c_i(y) - 1/2 \leq 0 , \quad y \in Y$$

with $c_1(y) = (2/y_1)\exp((-1/y_1)(1+(y_2-1)^2))$, for $y_1 > 0$

$c_2(y) = (1/2y_1)\exp((-1/2y_1)(1+y_2^2/4))$, for $y_1 > 0$

$c_3(y)$, $y_1 > 2$, and $c_{1,2,3}(y)$ elsewhere unchanged.

Again, due to the linearity of F and f with respect to x, z^k, k=0, can be chosen arbitrarily and (WSI) terminates after one iteration (k).

Starting point: $x^s = (0., 0., 1.)^T$, s=0.

Convergence - condition: relative accuracy 10^{-3} for z

	s=0	s=1	s=2	s=3, 4
x^s	0. 0. 1.	0.54261 0. 0.	0.54315 0. 0.12581	0.54315 0. 0.12615
$\overline{F}(x, z^{s-1}, z^k)$	—	1.08544	1.21252	1.21244

Execution time: about 4 sec.

There are two active points $y^{i,4}$: $y^{1,4} = (1.03890, 0.94190)^T$, $y^{2,4} = (2.98078, -0.85266)^T$ with Lagrange-multipliers $u^{1,4} = 2.15376$, $u^{2,4} = 2.77964$ respectively.

In the future more numerical experience must be gained with the proposed method. Also it is necessary to compare the method with other methods described in [4].

However the flexibility for changes in the number of active constraints, shown by the examples above, seems to be a improvement with regard to optimization methods used up to now.

References.

[1]. L. Cromme, Eine Klasse von Verfahren zur Ermittlung bester nichtlinearer Chebyshev - Approximationen, Numer. Math., 25 (1976), 447-459.

[2]. P.E. Gill and W. Murray, Numerically stable methods for quadratic programming, Math. Programming, 14 (1978), 349 - 372.

[3]. S.A. Gustafson and K.O. Kortanek, Mathematical models for air-pollution

control: determination of abatement policies, in: Models for Environmental Pollution Control, R. A. Deininger ed., Ann Arbor Science Press, Ann Arbor, Michigan, 1973, 251 - 265.

[4]. R. Hettich and W. van Honstede, On quadratically convergent methods for semi-infinite programming, this volume.

[5]. R. B. Wilson, A simplicial algorithm for concave programming, Graduate School of Business Administration, Harvard University, Cambridge, Mass., Dissertation, 1963.

ON SEMI-INFINITE PROGRAMMING IN NUMERICAL ANALYSIS

Sven-Ake Gustafson
Department of Numerical Analysis
Royal Institute of Technology
S-10044 Stockholm 70, Sweden

Abstract. In this paper we will illustrate how a large number of diverse numerical problems can be treated in a uniform manner by means of the theory and computational techniques of semi-infinite programming. From this point of view we shall discuss the construction of stable mechanical quadrature rules, nonlinear (in particular exponential) fitting to measured data and linear Čebyšev-approximation over multiple-dimensional sets.

1. Dual pairs of semi-infinite programs.

We introduce some notations which will be used throughout the paper: Let S be a fixed index-set, $a_1, a_2, \ldots, a_n$ and b $n+1$ functions defined on S. Let further $c \in R^n$ be a given vector. Then we define the two problems:

Program (P) Minimize the linear form

$$c^T y \tag{1a}$$

over all vectors $y \in R^n$ subject to the constraints

$$a(s)^T y \geq b(s), \quad s \in S \tag{1b}$$

and

Program (D) Determine the integer q, the subset $\{s_1, s_2, \ldots, s_q\} \subset S$ and the reals $x_1, x_2, \ldots, x_q$ such that the expression

$$\sum_{i=1}^{q} x_i b(s_i) \tag{2a}$$

is maximized under the constraints

(2b) $$\sum_{i=1}^{q} x_i a(s_i) = c$$

(2c) $$x_i \geq 0, \ i=1,2,\ldots,q \quad .$$

Here we have written $a(s)$ for the vector which has the components $a_r(s)$, $r=1,2,\ldots,n$. Superscript ^{T} denotes transposition, as usual.

If S is a finite set then (P) and (D) are linear programs, which may be solved by means of the simplex method. We know that when S is finite, then the conditions (1b) and (2b),(2c) may or may not be consistent and the preference functions (1a) and (2a) may or may not be bounded. Such exceptional cases will of cause give rise to computational difficulties if suitable precautions are not taken. The following result is often useful in the applications of semi-infinite programming (SIP).

Theorem 1.

a) Let program (P) meet *Slater's condition*, i.e. there is an $y^S \in R^n$ such that

$$a(s)^T y^S > b(s), \quad s \in S$$

where S is a compact subset of R^k.

b) There is a neighbourhood $N(c)$ to c such that all vectors $\tilde{c} \in N(c)$ admit a representation of the form (2b):

$$\sum_{i=1}^{\tilde{q}} \tilde{x}_i a(\tilde{s}_i) = \tilde{c}$$

$$\tilde{x}_i \geq 0 \qquad \tilde{s}_i \in S, \quad i=1,2,\ldots,\tilde{q}.$$

(The integer $\tilde{q}$ may depend on $\tilde{c}$).

Then we can state

i) The optimal values of programs (P) and (D) are equal (no duality gap).

(ii) Both programs assume their common optimal value; program (D) has an optimal solution with $q \leq n$ and such that the vectors $a(s_1), a(s_2),\ldots,a(s_q)$ are linearly independent.

iii) The optimal solutions of programs (P) and (D) satisfy the conditions

(3a) $\sum_{i=1}^{q} x_i a(s_i) = c$

(3b) $a(s_i)^T y = b(s_i)$, if $x_i > 0$, $i=1,2,\ldots,q$

(3c) $a^T y - b$ has a global minimum at s_i if $x_i>0$, $i=1,2,\ldots,q$

(3d) $x_i \geq 0$, $s_i \in S$, $i=1,2,\ldots,q$.

A proof of Theorem 1 can be assembled from the results in [9], Kap. III, IV.

One immediate consequence of Theorem 1 is, that if one is only interested in the optimal value there is a choice between solving program (P) or program (D). The latter may be looked upon as a conventional nonlinear constrained optimization problem. If the conditions of Theorem 1 are met, we can namely put $q=n$ from the outset since $q<n$ then corresponds to some of the x_i:s equal to zero.

In the applications of SIP it is fairly common that the index set S appears as the union of l disjoint subsets $S_1,S_2,\ldots,S_l$, each corresponding to a subset of conditions in(1b). We write out programs (P) and (D) explicitly for $l=2$ and obtain

Program (P_2): Minimize the linear form

$$c^T y$$

over all vectors $y \in R^n$ subject to the constraints

$$a(s)^T y \geq b(s),\ s \in S_1$$

and

$$a(s)^T y \geq b(s),\ s \in S_2$$

Program (D_2): Determine the integers q_1 and q_2 the subsets $\{s_1^1, s_2^1,\ldots,s_{q_1}^1\} \subset S_1$, $\{s_1^2,s_2^2,\ldots,s_{q_2}^2\} \subset S_2$ and the real numbers $x_1^1,x_2^1,\ldots,x_{q_1}^1$, $x_1^2,x_2^2,\ldots,x_{q_2}^2$ such that the expression

$$\sum_{i=1}^{q_1} x_i^{\,1} b(s_i^{\,1}) + \sum_{i=1}^{q_2} x_i^{\,2} b(s_i^{\,2})$$

is maximized under the constraints

$$\sum_{i=1}^{q_1} x_i^{\,1} a(s_i^1) + \sum_{i=1}^{q_2} x_i^{\,2} a(s_i^2) = c$$

and

$$x_i^{\,1} \geq 0,\ i=1,2,\ldots,q_1,\quad x_i^{\,2} \geq 0,\ i=1,2,\ldots,q_2\ .$$

Often some of the subsets $S_1, S_2, \ldots, S_1$ contain only finitely many elements. We illustrate this with the following task.

Minimize the linear form

(4a) $\quad c^T y$

over all vectors $y \in R^n$ subject to the constraints

(4b) $\quad a(s)^T y \geq b(s),\quad s \in S_1$

and

(4c) $\quad |y_r| \leq 1,\quad r=1,2,\ldots,n.$

An equivalent formulation of conditions (4b), (4c) is

$$a(s)^T y \geq b(s),\ s \in S_1$$

$$y_r \geq -1\ ,\ r=1,2,\ldots,n$$

$$-y_r \geq -1\ ,\ r=1,2,\ldots,n\ .$$

In this paper we shall use the convention that when we want to refer to an entire group of formulas which have the same number we indicate the number only. This we simply write (4) instead of (4a),(4b),(4c).

The problem defined by (4) is a particular instance of program (P) where S is represented as the union of three disjoint sets, two of which contain n elements each. For practical illustrations of the representation of S as a union of disjoint sets see e.g. [4],[8] and [10]. We mention in particular the air pollution abatement problem in [11] where one subset of conditions represents the constraints on the concentration in the air of the studied pollutant, another subset the constraints on the rate of fall-out on the ground.

Theorem 1 is often used for the computational solution of programs (P) and (D). Then conditions (3a),(3b), (3c) are combined to generate a nonlinear system of equations, involving finitely many variables, which is solved numerically. This idea is dicussed e.g. in [4],[7], [10],[12] and [13]. The major difficulties are then to determine q, find approximate values for $x_1,x_2,\dots,x_q$, $s_1,s_2,\dots,s_q$ and y with which to start iterative procedures, such as Newton-Raphson. One must also verify that the accepted s_i correspond to *global* minima of a^Ty-b. In some applications these obstacles are overcome by analytic considerations. To illustrate the nature of the general situation we discuss

Example 1. Minimize y_0 subject to

$$y_0 \geq b(s),\ s\in S$$

where b is a function defined on S. The solution is

$$y_0 = \sup_{s\in S} b(s)\ .$$

We note that it is impossible to estimate the error in a calculated value y_0 without quantitative information on the general behaviour of b, such as the value of a Lipschitz constant.

Thus in a general context one cannot hope to find an optimal solution but one must content oneself with determining the optimal value within a prescibed error bound. The following lemma sometimes called the *weak duality lemma* is often used to estimate the error in calculated optimal values.

Lemma 1. Let y meet (1b),q, $\{s_1,s_2,\dots,s_q\}$ and $x_1,x_2,\dots,x_q$ (2b), (2c). Then

$$\sum_{i=1}^{q} x_i b(s_i) \leq c^Ty$$

Proof.

$$c^Ty = y^Tc = y^T \sum_{i=1}^{q} x_i a(s_i) = \sum_{i=1}^{q} x_i a(s_i)^T y \geq \sum_{i=1}^{q} x_i b(s_i) \quad ||.$$

Due to Lemma 1 we will seek "good" feasible solutions. A systematic method of doing this is by means of *discretization:* We select a finite subset $T \subset S$ and replace S with T in programs (P) and (D). The linear programs hereby arising are solved numerically. If the assumptions of Theorem 1 prevail, S is compact and $a_1, a_2, \ldots, a_n, b$ are continuous on S, then the discretization error may be expressed in terms of

$$|T| = \max_{s \in S} \min_{t \in T} \operatorname{dist}(s,t)$$

where dist is a distance function defined on S. See [4] and [10]. A general method to solve SIP is first to treat a discretized version of programs (P) and (D), by means of stabilized linear programming. Based on the optimal solutions of these discretized programs, a non-linear system is formed by combining (3a), (3b) and (3c). This system is solved numerically. Efficient computer codes implementing this scheme are given in [3]. For further details see e.g. [4],[10].

2. One-sided approximation and stable quadrature rules.

We now introduce

Assumption CCLI: S is Compact and the functions $a_1, a_2, \ldots, a_n$ and b are Continuous and Linearly Independent on S.

Let L be a positive bounded linear functional on $C(S)$, the linear space of functions, continuous on the compact set S. Then $L(f)$ has a representation

(5) $L(f) = \int f(s) d\alpha(s), \quad f \in C(S),$

where $d\alpha$ is a bounded nonnegative measure independent of f. Put $c_r = L(a_r)$, $r = 1,2,\ldots,n$. Then program (P) takes the form

(6a) Minimize $\int_S y^T a(s) d\alpha(s)$

(6b) subject to $y^T a(s) \geq b(s), \; s \in S.$

One verifies directly that if y^1 is an optimal solution of the problem defined by (6) then y^1 is also an optimal solution of the task to minimize

(7) $\int_S |y^T a(s) - b(s)| \; d\alpha(s)$

subject to (6b). The expression (7) is the error in the weighted L_1-norm defined by $d\alpha$ when we approximate b from the above by the linear combination

$$\sum_{r=1}^{n} y_r a_r \quad .$$

In the same way we may determine the best approximation from the below by maximizing the integral in (6a) subject to

$$y^T a(s) \le b(s), \quad s \in S \ .$$

Let y^1 and y^2 be the best approximations from the above and below respectively. Then one may expect that $y = \hat{y}$ with

$$\hat{y} = \frac{1}{2}\,[y^1 + y^2]$$

should give a small value of

$$\int_S |y^T a(s) - b(s)|\, d\alpha(s)$$

and y has been calculated without access to $d\alpha(s)$, $s \in S$ since only the vector $c \in R^n$ was required. We note here that under assumption CCLI and the assumptions of Theorem 1, arbitrarily close approximations to y^1 and y^2 may be computed using discretization and linear programming.

We next consider program (D) and let $q, \{s_1, s_2, \ldots, s_q\}$, $x_1, x_2, \ldots, x_q$ meet (2b),(2c). Put for arbitrary $z \in R^n$

$$Q(s) = z^T a \quad .$$

Then we find

$$\sum_{i=1}^{q} x_i Q(s_i) = \sum_{i=1}^{q} x_i z^T a(s_i) = c^T z = L(Q) \ .$$

Thus we have a mechanical quadrature rule with nonnegative weights which evaluates the Stieltjes integral (5) exactly for all linear combinations of $a_1, a_2, \ldots, a_n$.

We next relax the condition that L is a positive functional but still seek to determine quadrature rules by treating problems similar to program (D). We minimize the sum of the absolute values of the weights and hence arrive at the problem:

Determine the integer q, the subset $\{s_1, s_2, \ldots, s_q\} \subset S$ and the reals

$x_1, x_2, \dots, x_q$ such that the expression

(8a) $\quad - \sum_{i=1}^{q} |x_i|$ is maximized under the constraints

(8b) $\quad \sum_{i=1}^{q} x_i a(s_i) = c$.

Since $c_r = La_r$, $r = 1,2,\dots,n$ the relation (8b) implies that the quadrature rule

$$\sum_{i=1}^{q} x_i f(s_i) \approx LF$$

gives exact results if f is a linear combination of $a_1, a_2, \dots, a_n$. It is always possible to determine q, $x_1, x_2, \dots, x_q$ and $s_1, s_2, \dots, s_q$ such that (8b) is satisfied if the functions $a_1, a_2, \dots, a_n$ are linearly independent. Because then there are n elements $t_i \in S$ such that the vectors $a(t_i)$, $i = 1,2,\dots,n$ are linearly independent. The fact that the problem (8) can be solved by means of SIP is a consequence of

Lemma 2. The problem (8) is equivalent to the following task:

Determine the integers q^+ and q^-, the subsets $S^+ = \{s_1^+, s_2^+, \dots, s_{q^+}^+\} \subset S$ $S^- = \{s_1^-, s_2^-, \dots, s_{q^-}^-\} \subset S$ and the reals $x_1^+, x_2^+, \dots, x_{q^+}^+$, $x_1^-, x_2^-, \dots, x_{q^-}^-$ such that the expression

(9a) $\quad - \sum_{i=1}^{q^+} x_i^+ - \sum_{i=1}^{q^-} x_i^-$ is maximized

under the constraints

(9b) $\quad \sum_{i=1}^{q^+} x_i^+ a(s_i^+) - \sum_{i=1}^{q^-} x_i^- a(s_i^-) = c$

(9c) $\quad x_i^+ \geq 0 \quad i = 1,2,\dots,q^+ , \quad x_i^- \geq 0, \ i = 1,2,\dots,q^-$

Proof. We note that the optimum value of (9a) is not changed if we introduce the additional constraint

(9d) $\quad s_i^+ = s_j^- \Rightarrow x_i^+ \cdot x_j^- = 0.$

Assume namely that

$$s_i^+ = s_j^-, \ \min(x_i^+, x_j^-) = d > 0.$$

If we replace x_i^+ by $x_i^+ - d$ and x_j^- by $x_j^- - d$, then (9b), (9c) are still satisfied but the value of (9a) increases by $2 \cdot d > 0$. Thus we assume that (9b), (9c) and (9d) are all satisfied when we establish the claimed equivalence. Let now q, $\{s_1, s_2, \ldots, s_q\} \subset S$ and $x_1, x_2, \ldots, x_q$ satisfy (8b) and denote by v the corresponding value of (8a). Put now

$$x_i^+ = \frac{1}{2}(|x_i| + x_i) \quad x_i^- = \frac{1}{2}(|x_i| - x_i), \quad i = 1,2,\ldots,q,$$

and

$$S^+ = \{s_i | x_i^+ > 0\} \quad S^- = \{s_i | x_i^- > 0\}$$

Then S^+ and S^- are disjoint. Putting q^+ equal to the number of elements in S^+ and q^- equal to the number of elements in S^- we find that (9b), (9c), (9d) are satisfied and that (9a) also has the value v. Asumme now that the integers q^+, q^-, the subsets $S^+ = \{s_1^+, s_2^+, \ldots, s^+_{q^-}\} \subset S$, $S^- = \{s_1^-, s_2^-, \ldots, s^-_{q^-}\} \subset S$, and the reals $x_1^+, x_2^+, \ldots, x^-_{q^+}$, $x_1^-, x_2^-, \ldots, x^-_{q^-}$ satisfy (9b),(9c),(9d). Let w be the corresponding value of the preference function (9a). By (9d) S^+ and S^- are disjoint sets. Put $S = S^+ \cup S^-$, $q = q^+ + q^-$. Write $S = \{s_1, s_2, \ldots, s_q\}$. If $s_\ell = s_i \in S^+$ we put $x_\ell = x_i^+$ and if $s_\ell = s_j \in S^-$ we put $x_\ell = -x_j^-$. Hence (8b) is satisfied and (8a) assumes also the value w establishing the assertion.

We now prove

<u>Theorem 2.</u> Let assumption CCLI prevail. Then there is a quadrature rule which solves the program (8) and is such that $q \le n$.

<u>Proof.</u> By Lemma 2 (8) is equivalent to the task defined by (9a), (9b), (9c). (we exercise our option to exclude (9d)). Using the definition of the dual pair (P) and (D) we find that (9a),(9b),(9c) is the dual of the problem

(10a) Minimize $y^T c$

under the constraints

(10b) $|a(s)^T y| \le 1$, $s \in S$.

The vector $y = 0$ meets the Slater condition. It has already been pointed out that (8b) and hence (9b), (9c) can be satisfied for all $c \in R^n$. Therefore the desired result follows from Theorem 1. ||

In a practical situation it is often not necessary to solve program (8) exactly but an approximate solution is enough. Then one may proceed in analogy with the suggestion at the end of Section 1: The set S in programs (8) and (9) is replaced with a finite subset and the optimization problems hereby arising are solved by means of linear programming.

3. A class of nonlinear approximation problems.

In this Section we discuss a general class of nonlinear approximation and fitting problems which contains the task to fit an exponential sum to measured data as an important special case.

Let $g(t,s)$ be a function of two variables, which is defined for $t \geq 0$ and $s \in [0,1]$. Let further $f(t)$ be defined for $t \geq 0$. We want to approximate f with a sum

(11) $$f^*(t) = \sum_{i=1}^{q} x_i g(t,s_i)$$

where $x_i \geq 0$, $i = 1,2,...,q$ and $0 \leq s_1 < s_2 < ... < s_q$.

We get exponential approximation as a special instance, if we take $g(t,s) = s^t$. Then we obtain

$$f^*(t) = \sum_{i=1}^{q} x_i s_i^{\,t} = \sum_{i=1}^{q} x_i e^{-\lambda_i t} \quad \text{with} \quad e^{-\lambda_i} = s_i$$

Let n be given and $0 = t_1 < t_2 < ... < t_n$ be fixed numbers. We consider the two approaches i) and ii) below to determine $q, s_1, s_2, ..., s_q$ and $x_1, x_2, ..., x_q$

i) Interpolation. We require f^* defined by (11) to *interpolate* f at $t_1, t_2, ..., t_n$. Then q, $s_1, s_2, ..., s_q$ and $x_1, x_2, ..., x_q$ shall meet the conditions

(12a) $$\sum_{i=1}^{q} x_i g(t_r, s_i) = f(t_r), \quad r = 1,2,...,n$$

(12b) $$x_i \geq 0 \quad i = 1,2,...,q$$

(12c) $$0 \leq s_1 < s_2 < ... < s_q \leq 1.$$

If the conditions (12a), (12b), (12c) are consistent, then $q, s_1, s_2, ..., s_q$ and $x_1, x_2, ..., x_q$ may be determined in infinitely many ways. It is therefore advantageous to inpose further conditions. Let b be defined on $[0,1]$. We then discuss the problem to

(12d) maximize $\sum_{i=1}^{q} x_i b(s_i)$

subject to the constraints (12a), (12b), (12c). We immediately recognize this task as a special instance of program (D). (The requirement $s_1 < s_2 < \ldots < s_q$ is met by numbering s_i appropriately).

Let now t be a fixed number and put $b(s) = g(t,s)$ in (12d). Then the problem defined by (12) is the task to determine the maximum value of $f^*(t)$ over all sums f^* of the form (11) which interpolate f at $t_1,t_2,\ldots,t_n$. The case when the functions $f(t_r,\cdot)$ $r = 1,2,\ldots,n$ and b form a Čebyšev system is discussed in [14]. The numerical treatment may be carried out according to [5] but one may also use the general procedure described in Section 1. A particular instance of this case occurs when we take $g(t_r,s) = s^{r-1}$ $b(s) = (1+s)^{-1}$. It appears when one seeks the sum of certain slowly converging series. See [2]

ii) Approximation: We require f^* defined by (11) to *approximate* f at the points $t_1,t_2,\ldots,t_n$. We define the vector $\delta \in R^n$ through

$$f^*(t_r) + \delta_r = f(t_r), \quad r = 1,2,\ldots,n .$$

Then we introduce a norm on R^n and consider the problem to minimize $||\delta||$ over all f^* defined by (11) and subject to the constraints (12b), (12c). The case

$$||\delta||^2 = \sum_{r=1}^{n} \delta_r^2$$

is discussed in [15]. Here we will consider the choice

$$||\delta|| = \sum_{r=1}^{n} |\delta_r|$$

which can be treated by means of SIP. We thus consider the task

(13a) Maximize $-\sum_{r=1}^{n} |\delta_r|$

subject to the constraints

(13b) $\sum_{i=1}^{q} x_i g(t_r,s_i) + \delta_r = f(t_r), \quad r=1,2,\ldots,n .$

(13c) $x_i \geq 0, \quad i = 1,2,\ldots,q .$

(13d) $\quad 0 \le s_1 < s_2 < \ldots < s_q \le 1$

Arguing as in the proof of Lemma 2 we conclude that program (13) is equivalent to the problem

(14a) $\quad$ Maximize $\quad -\sum_{r=1}^{n} (\delta_r^+ + \delta_r^-)$

subject to the constraints

(14b) $\quad \sum_{i=1}^{q} x_i g(t_r, s_i) + \delta_r^+ - \delta_r^- = f(t_r), \; r = 1,2,\ldots,n$

(14c) $\quad x_i \ge 0, \; i = 1,2,\ldots,q, \quad \delta_r^+ \ge 0, \; \delta_r^- \ge 0, \quad r=1,2,\ldots,n$

(14d) $\quad 0 \le s_1 < s_2 < \ldots < s_q \le 1.$

We observe that program (14) is an instance of program (D). It is consistent since

$$q=0 \quad \delta_r^+ = \tfrac{1}{2}\{|f(t_r)| + f(t_r)\} \; \delta_r^- = \tfrac{1}{2}\{|f(t_r)| - f(t_r)\}$$

trivially meets the conditions (14b), (14c), (14d). Approximations to the optimal value can be found by means of discretization of [0,1] and linear programming.

Remark. Other variants of the problems discussed in this section may be of interest. Thus one could treat the task

(15a) $\quad$ Maximize $\quad \sum_{i=1}^{q} x_i b(s_i)$

subject to the constraints

(15b) $\quad \sum_{i=1}^{q} x_i g(t_r, s_i) + \delta_r = f(t_r), r=1,2,\ldots,n.$

(15c) $\quad x_i \ge 0, \; i=1,2,\ldots,q \;,$

(15d) $\quad 0 \le s_1 < s_2 < \ldots < s_q \le 1,$

(15e) $\quad ||\delta|| \le \delta_0 \;.$

Thus problem occurs when one wants to maximize the functional (15a) and the values $f(t_r)$ are beset with measurement errors whose norm is bounded by δ_0. This situation is conceivable when one wants to make predictions based on measured quantities.

4. Application of the weak duality lemma to two-sided approximation

Let $a_1, a_2, \ldots, a_n$, b and S meet Assumption CCLI of Section 2. We want to approximate b by a linear combination of $a_1, a_2, \ldots, a_n$ in such a manner that the error measured in the maximum norm is rendered a minimum. This we seek

$$y_0 = \min_{y \in R^n} \max_{s \in S} |b(s) - y^T a(s)| .$$

This task may also be written

(16a) Minimize y_0

over all $y \in R^n$ subject to the constraints

(16b) $|y^T a(s) - b(s)| \leq y_0, \quad s \in S$.

An equivalent formulation is

(17a) Minimize y_0

over all $y \in R^n$ subject to the constraints

(17b) $y^T a(s) + y_0 \geq b(s), \quad s \in S$

(17c) $-y^T a(s) + y_0 \geq -b(s), \quad s \in S$

(17d) $y_0 \geq 0$

Program (17) is clearly an instance of program (P) of Section 1. Its dual reads: Determine the integers $q+$ and q^-, the subsets $\{s_1^+, s_2^+, \ldots, s_{q^+}^+\} \subset S$, $\{s_1^-, s_2^-, \ldots, s_{q^-}^-\} \subset S$ and the reals $x_1^+, x_2^+, \ldots, x_{q^+}^+$, $x_1^-, x_2^-, \ldots, x_{q^-}^-$ such that the expression

$$\sum_{i=1}^{q^+} x_i^+ b(s_i^+) - \sum_{i=1}^{q^-} x_i^- b(s_i^-)$$

is maximized under the constraints

$$\sum_{i=1}^{q^+} x_i^+ a(s_i^+) - \sum_{i=1}^{q^-} x_i^- a(s_i^-) = 0$$

$$\sum_{i=1}^{q^+} x_i^+ + \sum_{i=1}^{q^-} x_i^- + x_0 = 1$$

$x_i^{\mp} \geq 0$, $i=1,2,\ldots,q^+$, $x_i^- \geq 0$, $i=1,2,\ldots,q^-$, $x_0 \geq 0$.

One easily proves that an equivalent formulation is

(18a) Minimize $\sum_{i=1}^{q} x_i b(s_i)$

over all integers q, subsets $\{s_1,s_2,\ldots,s_q\} \subset S$ and reals $x_1,x_2,\ldots,x_q$ under the constraints

(18b) $\sum_{i=1}^{q} x_i a(s_i) = 0$

(18c) $\sum_{i=1}^{q} |x_i| = 1$.

Thus the problems (16) and (18) are a dual pair. By Theorem 1 they assume their common optimal value. They can be solved by deriving a nonlinear system of equations from complementary slackness and solving it numerically. Since this approach is discussed extensively elsewhere (see e.g. [1],[12] and [13]) we shall not pursue it here but instead discuss how approximate solutions are obtained by means of elementary methods.

The weak duality lemma applied to (16) and (18) takes the form

Lemma 3. Let the integer q, the subset $\{s_1,s_2,\ldots,s_q\} \subset S$ and the reals $x_1,x_2,\ldots,x_q$ satisfy (18b), (18c). Let further $y \in R^n$ be a fixed vector and put

(19) $y_0 = \max_s |b(s) - y^T a(s)|$

Then

(20) $|\sum_{i=1}^{q} x_i b(s_i)| \leq y_0$ ||

Remark. Put

(21) $Q(s) = y^T a(s)$

Since

$$\sum_{i=1}^{q} x_i\{b(s_i) - Q(s_i)\} = \sum_{i=1}^{q} x_i b(s_i) - y^T \sum_{i=1}^{q} x_i a(s_i) = \sum_{i=1}^{q} x_i b(s_i)$$

we may combine (19) and (20) into

$$(22) \quad |\sum_{i=1}^{q} x_i\{b(s_i) - Q(s_i)\}| \le v \le \max_{s\in S}|b(s)-Q(s)|$$

where v is the joint optimal value of programs (16) and (18). Therefore if a linear combination Q, (21) and the associated deviation (19) is known, it is possible to calculate a lower bound for the deviation corresponding to an optimal solution. One may hence assess the improvements in the approximation that will result from the effort to calculate an optimal solution.

Good approximations are calculated systematically as follows: The set S is replaced by a finite subset T. Using linear programming we solve problems (16) and (18) when T has replaced S in these optimization problems. In the process we have got $q, s_1, s_2, \ldots, s_q$ and $x_1, x_2, \ldots, x_q$ satisfying (18b) and (18c). Let now $\tilde{y}$ be an optimal solution of the discretized version of (16) and $\tilde{y}_0$ the corresponding optimal value. Thus the left hand side of (22) has the value $\tilde{y}_0$ if we put $Q = \tilde{y}^T a$ i.e. the largest value of $|y^T a(s) - b(s)|$ at the *grid* and the right-hand side of (22) indicates how far $Q(s)$ deviates from $b(s)$ *between* the gridpoints.

In the special case $S = [-1,1]$, $a_r(s) = s^{r-1}$ it is well-known that a good approximation is obtained by interpolating b at the points $\cos \frac{i-1/2}{n}\pi$, $i=1,2,\ldots,n$. Let Q be the corresponding interpolating polynomial. Since in this case (18b), (18c) are satisfied by $q = n+1$, $s_i = \cos \frac{(i-1)\pi}{n}$, $i=1,2,\ldots,n+1$, $|x_1| = |x_{n+1}| = \frac{0.5}{n}$, $|x_i| = 1/n$, $i=1,2,\ldots,n$ the inclusion (22) takes the special form

$$|\frac{1}{n} \sum_{i=1}^{n+1}{}'' (-1)^i b(\cos \frac{(i-1)\pi}{n})| \le v \le \max_{s}|b(s)-Q(s)|.$$

where $''$ as usual means that the first and the last term in the sum shall be premultiplied by the factor $1/2$.

References

1. Andreassen, D.O. and C.A. Watson, Linear Chebyshev approximation without Chebyshev sets, BIT 16 (1976), 349-362.

2. Dahlquist, G.,S.-Å. Gustafson and K. Siklo'si, Convergence acceleration from the point of view of linear programming, BIT 5 (1965), 1-16.

3. Fahlander, K., Computer programs for semi-infinite optimization, TRITA-NA-7312, Dept. of Numerical Analysis, Royal Institute of Technology, S-10044 Stockholm 70, Sweden, 1973.

4. Gustafson, S.-Å., On the computational solution of a class of generalized moment problems, SIAM J. on Numer. Anal. 7 (1970), 343-357.

5. Gustafson, S.-Å., Die Berechnung von verallgemeinerten Quadraturformeln vom Gaußschen Typus, eine Optimierungsaufgabe,in L. Collatz, W. Wetterling (eds.): Numerische Methoden bei Optimierungsaufgaben, ISNM 17 (1973), 59-71, Birkhäuser, Basel.

6. Gustafson, S.-Å., On computational applications of the theory of moment problems, Rocky Mountain J. on Math. 4 (1974), 227-240.

7. Gustafson, S.-Å., Nonlinear systems in semiinfinite programming, in G.B. Byrnes, C.A. Hall (eds.), Numerical Solution of Nonlinear Algebraic Systems, Academic Press, 1973, pp. 63-99.

8. Gustafson, S.-Å., Some optimization problems in numerical analysis, Methods of Operations Research, 25 (1977), 367-379.

9. Glashoff,K. and Gustafson,S.-Å, Einführung in die lineare Optimierung, Wissenschaftliche Buchgesellschaft, Darmstadt, 1978.

10. Gustafson, S.-Å. and K. Kortanek, Numerical treatment of a class of semiinfinite programming problems, NRLQ 20 (1973), 477-504.

11. Gustafson, S.-Å. and K. Kortanek, On the calculation of optimal long-term air pollution abatement strategies for multiple-source areas II, Proceedings of the Sixth Meeting of the NATO/CCMS expert panel on air pollution modeling, 1976.

12. Hettich, R., Kriterien zweiter Ordnung für lokal beste Approximation, Numer. Math. 22 (1974), 409-417.

13. Hettich,R., A Newton-method for nonlinear Chebyshev approximation, in Approximation Theory, Proc. Int. Colloquium Bonn 1976, Springer Lecture Notes on Math. No. 556 (1976), 222-236.

14. Karlin, S. and W.J. Studden: Tchebycheff Systems: with Applications in Analysis and Statistics, Interscience Publishers, J. Wiley and Sons, 1966.

15. Ruhe, A., Least squares fitting by positive sums of exponentials, UMIMF-70.78, Institute of Information Processing, Univ. of Umeå, S-90187 Umeå, Sweden.

BOUNDS FOR THE ERROR IN LINEAR SYSTEMS*

Germund Dahlquist, Gene H. Golub, and Stephen G. Nash.

1. INTRODUCTION.

Consider the system of linear algebraic equations

$$A\underset{\sim}{x} = \underset{\sim}{b} \tag{1.1}$$

where A is a real, symmetric positive definite n x n matrix and $\underset{\sim}{b}$ is a given vector. Assume that we have an approximate value for $\underset{\sim}{x}$, say $\underset{\sim}{\xi}$, so that

$$\underset{\sim}{x} = \underset{\sim}{\xi} + \underset{\sim}{e} \tag{1.2}$$

where $\underset{\sim}{e}$ is denoted as the error vector. We wish to determine upper and lower bounds for $\|\underset{\sim}{e}\|$ where $\|\cdot\|$ indicates the euclidean norm of the vector.

In order to compute bounds for the norm of the error vector, it is natural to compute the residual vector,

$$\underset{\sim}{r}_0 = \underset{\sim}{b} - A\underset{\sim}{\xi} \quad . \tag{1.3}$$

Thus, since $\underset{\sim}{r}_0 = A\underset{\sim}{e}$,

$$\frac{\|\underset{\sim}{r}_0\|}{\|A\|} \le \|\underset{\sim}{e}\| \le \|A^{-1}\| \, \|\underset{\sim}{r}_0\| .$$

Here $\|A\|$ indicates the spectral norm of A. Assuming that $\|A\|=1$ (this can be accomplished via a simple scaling of (1.1)), we see that even though $\|\underset{\sim}{r}_0\|$ is "small", the upper bound for $\|\underset{\sim}{e}\|$ can be quite large when $\|A^{-1}\|$ is large.

By computing additional information, it is possible to obtain more precise upper and lower bounds on the euclidean norm of the vector. In section 2, we give an algorithm which requires an auxiliary sequence of vectors and an explicit knowledge of all the eigenvalues of the matrix A. The bounds are then obtained as a solution to a linear programming problem.

In section 3, the same auxiliary sequence of vectors is used to estimate the error but it is only assumed that an upper bound on the largest eigenvalue and a positive lower bound on the smallest eigenvalue are known. Using the theory of moments, an algorithm is given for determining upper and lower bounds for $\|\underset{\sim}{e}\|$. In section 4, we show how the Lanczos algorithm may be used for computing the bounds. In section 5, we show how an improved solution can be obtained from the conjugate gradient method and bounds for the error are given for the improved solution. Finally, numerical experiments are described in section 6.

The results in sections 2, 3 and 4 were originally described elsewhere ([3], [8]) but those in section 5 appear to be new.

---

*This research was supported in part by Department of Energy contract EY-76-S-03-0326 PA # 30.

Throughout this paper, vectors will be denoted by underlining (e.g. $\underset{\sim}{a}$, $\underset{\sim}{b}$,...) and the (i,j) entry of a matrix A will be indicated by a_{ij} or $\{A\}_{ij}$.

2. BOUNDS USING LINEAR PROGRAMMING.

Consider the Krylov sequence

$$\underset{\sim}{r}_{i+1} = A\,\underset{\sim}{r}_i \qquad (i = 0,1,\dots,k-1)$$

where $\underset{\sim}{r}_0$ is defined by (1.3). Thus

$$\underset{\sim}{r}_i = A^i\underset{\sim}{r}_0 \qquad (i = 0,1,\dots,k) .$$

We define

$$(\underset{\sim}{x},\underset{\sim}{y}) \equiv \sum_{i=1}^{n} x_i y_i$$

so that

$$\begin{aligned}(\underset{\sim}{r}_p,\underset{\sim}{r}_q) &= (A^p\underset{\sim}{r}_0,\, A^q\underset{\sim}{r}_0)\\ &= (A^{p+q}\underset{\sim}{r}_0,\, \underset{\sim}{r}_0)\\ &\equiv \mu_{p+q} \qquad (p,q = 0,1,\dots,k) .\end{aligned}$$

Since A is symmetric and positive definite, we have

$$A\,\underset{\sim}{u}_i = \lambda_i\,\underset{\sim}{u}_i \qquad (i = 1,2,\dots,k)$$

with

$$(\underset{\sim}{u}_i,\,\underset{\sim}{u}_j) = \begin{cases} 0 & \text{for } i \neq j \\ 1 & \text{for } i = j \end{cases}$$

and

$$0 < a \leq \lambda_1 \leq \dots \leq \lambda_n \leq b .$$

Writing

$$\underset{\sim}{r}_0 = \sum_{i=1}^{n} \alpha_i\,\underset{\sim}{u}_i ,$$

we have

$$\begin{aligned}\mu_m &= (A^m\,\underset{\sim}{r}_0,\,\underset{\sim}{r}_0)\\ &= \sum_{i=1}^{n} \alpha_i^2\,\lambda_i^m \qquad (m = 0,1,\dots,2k). \qquad (2.1)\end{aligned}$$

Since $\underset{\sim}{e} = A^{-1}\,\underset{\sim}{r}_0$,

$$\|\underset{\sim}{e}\|^2 = (A^{-2}\underset{\sim}{r}_0,\,\underset{\sim}{r}_0) = \sum_{i=1}^{n} \alpha_i^2\,\lambda_i^{-2} = \mu_{-2} \qquad (2.2)$$

Equations (2.1) and (2.2) are equivalent to

$$\mu_m = \int_a^b \lambda^m\,d\,\alpha(\lambda) \qquad (m = -2,0,1,\dots,2k) \quad (2.3)$$

where

$$\alpha(\lambda) = 0 \qquad \text{for} \qquad a \leq \lambda < \lambda_1$$

$$\alpha(\lambda) = \sum_{i=1}^{r} \alpha_i^2 \qquad \text{for} \qquad \lambda_r \leq \lambda < \lambda_{r+1} \qquad (r = 1,2,\ldots,n-1)$$

$$\alpha(\lambda) = \sum_{i=1}^{n} \alpha_i^2 \qquad \text{for} \qquad \lambda_n \leq \lambda \leq b$$

The problem of determining and upper and lower bound for $\|\underset{\sim}{e}\|$ is equivalent to the following.

Problem 1

Given the $(2k+1)$ moments μ_i, determine upper and lower bounds for μ_{-2}. The solution of this classical problem (cf.[11]) is dependent upon the known information.

Suppose the eigenvalues of $A, \{\lambda_i\}_{i=1}^n$, are known but the coefficients $\{\alpha_i^2\}_{i=1}^n$ are unknown. Then to determine an upper bound for $\|\underset{\sim}{e}\|^2$, we find the maximum of

$$\sum_{i=1}^{n} \gamma_i \lambda_i^{-2}$$

subject to the constraints

$$\sum_{i=1}^{n} \gamma_i \lambda_i^m = \mu_m \qquad (m = 0,1,\ldots,2k)$$

$$\gamma_i \geq 0 \qquad (i = 1,2,\ldots,n).$$

The numerical solution of this problem can be obtained by the Simplex Algorithm of G. Dantzig [4]. Special techniques may be used to take advantage of the fact that a Vandermonde system is solved at each iteration.

Now let us assume that we are given a and b such that

$$0 < a \leq \lambda_i \leq b \qquad (i = 1,2,\ldots,n)$$

We want to determine an upper bound (U) and lower bound (L) such that

$$L \leq \int \varphi(\lambda)\, d\,\alpha(\lambda) \leq U$$

where $\varphi(\lambda)$ is a given function and we are given

$$\mu_i = \int \lambda^i\, d\,\alpha(\lambda) \qquad (i = 0,1,\ldots,2k).$$

Suppose we construct a polynomial

$$\pi_{2k}(\lambda) \equiv c_0 + c_1\lambda + \ldots. + c_{2k}\lambda^{2k}$$

such that

$$\pi_{2k}(\lambda) \geq \varphi(\lambda) \qquad a \leq \lambda \leq b\,.$$

Now $\pi_{2k}(\lambda)$ is not unique but note that

$$\int \pi_{2k}(\lambda)\, d\,\alpha(\lambda) = c_0\mu_0 + \ldots. + c_{2k}\mu_{2k}\,.$$

Hence we seek that polynomial $\hat{\pi}_{2k}(\lambda)$ such that

$$\hat{\pi}_{2k}(\lambda) \geq \varphi(\lambda) \qquad a \leq \lambda \leq b$$

and

$$\sum_{i=0}^{2k} \hat{c}_i \mu_i \leq \sum_{i=0}^{2k} c_i \mu_i \quad .$$

For a general function $\varphi(\lambda)$, in order to compute $\hat{\pi}(\lambda)$ one uses semi-infinite programming. Since in our situation $\varphi(\lambda) = \lambda^s$, it is possible to explicitly solve for L and U using the theory of moments.

3. ERROR BOUNDS USING THE THEORY OF MOMENTS.

The problem of determining bounds on μ_{-2} is related to the classical theory of moments which has been developed by A. A. Markov. In order to give a numerical algorithm for determining bounds for $\|e\|$, we review some facts from the theory of Gaussian quadrature.

Suppose we are given $\{\mu_i\}_{i=0}^{2k}$, and a function $\varphi(\lambda)$ $(a \leq b)$, and we wish to determine (L, U) so that

$$L \leq \int_a^b \varphi(\lambda) d\, \alpha(\lambda) \leq U$$

We can determine a quadrature rule such that

$$\mu_r = \int_a^b \lambda^r d\, \alpha(\lambda) = \sum_{i=1}^{k} A_i t_i^r + \sum_{j=1}^{m} B_j z_j^r \qquad \text{for } r = 0,1,\ldots,2k + m - 1 ,$$

where $\{A_i, t_i\}_{i=1}^{k}$ and $\{B_j\}_{j=1}^{m}$ are unknown and $\{z_j\}_{j=1}^{m}$ is specified. Then

$$\int_a^b \varphi(\lambda) d\, \alpha(\gamma) = \sum_{i=1}^{k} A_i \varphi(t_i) + \sum_{j=1}^{m} B_j \varphi(z_j) + R[\varphi] ,$$

where

$$R[\varphi] = \frac{\varphi^{(2k+m)}(\eta)}{(2k+m)!} \int_a^b \prod_{j=1}^{m} (\lambda - z_j) \left[\prod_{i=1}^{k} (\lambda - t_i) \right]^2 d\alpha(\lambda) \qquad a < \eta < b . \tag{3.1}$$

Thus if $\varphi(\lambda) = \lambda^{-2}$ and $m = 1$,

$$R[\lambda^{-2}] = -2(k+1)\eta^{-(2k+3)} \int_a^b (\lambda - z_1) \left[\prod_{i=1}^{k} (\lambda - t_i) \right]^2 d\alpha(\lambda) .$$

Hence for $z_1 = a > 0$, the Gauss-Radau type quadrature rule yields an upper bound for $\int_a^b \lambda^{-2} d\alpha(\lambda)$ and if $z_1 = b$, we have a lower bound. It can be shown (cf.[1, p. 80]) that these bounds are attainable.

Unfortunately, using the moments for computing the quadrature rules is a very ill-conditioned numerical problem [6]. We can avoid this difficulty by working with orthogonal polynomials which are defined by the distribution function $\alpha(\lambda)$.

4. THE LANCZOS ALGORITHM.

Associated with the distribution function $\alpha(\lambda)$, there is a set of orthogonal polynomials, namely $\{p_j(\lambda)\}$ with

$$\int p_j(\lambda)\, p_k(\lambda)\, d\alpha(\lambda) = 0 .$$

It is well known that these polynomials satisfy the recurrence relationship

$$p_{j+1}(\lambda) = (\lambda - \xi_{j+1})\, p_j(\lambda) - \eta_j^2\, p_{j-1}(\lambda) \tag{4.1}$$

with $p_{-1}(\lambda) = 0$, $p_0(\lambda) = 1$. The zeroes of $p_k(\lambda)$ are the nodes of the Gauss quadrature rule associated with $\alpha(\lambda)$. Thus

$$p_k(t_i) = 0 \qquad (i = 1,2,..,k) \ ,$$

and

$$\mu_r = \int_a^b \lambda^r d\alpha(\lambda) = \sum_{i=1}^{k} A_i t_i^r \qquad (r = 0,1,..,2k-1).$$

The coefficients $\{\xi_j\}_{j=1}^{k}$, $\{\eta_j^2\}_{j=1}^{k-1}$ can be computed from the moments but this process is numerically unstable (cf.[6]). The polynomials $\{\tilde{p}_j(\lambda)\}$ generated by the recurrence relationship

$$\eta_{j+1}\, \tilde{p}_{j+1}(\lambda) = (\lambda - \xi_{j+1})\, \tilde{p}_j(\lambda) - \eta_j\, \tilde{p}_{j-1}(\lambda) \tag{4.2}$$

will be orthonormal on the interval (a,b).

The coefficients of the recurrence relationship (4.1) can be computed directly by using the Lanczos algorithm [12]. We generate a sequence of vectors $\{\underset{\sim}{z}_j\}_{j=0}^{k}$ such that

$$\underset{\sim}{z}_i^T \underset{\sim}{z}_j = \begin{cases} 0 & \text{for } i \neq j \\ 1 & \text{for } i = j \end{cases}$$

Define

$$\underset{\sim}{z}_0 = \underset{\sim}{r}_0 \times (\|\underset{\sim}{r}_0\|)^{-1} \ .$$

Then for

$$\left.\begin{aligned} j &= 0,1,\dots,k, \\ \xi_{j+1} &= \underset{\sim}{z}_j^T A \underset{\sim}{z}_j \\ \underset{\sim}{w}_j &= A \underset{\sim}{z}_j - \xi_{j+1}\underset{\sim}{z}_j - \eta_j \underset{\sim}{z}_{j-1} \qquad (\eta_0 = 0) \\ \eta_{j+1} &= \|\underset{\sim}{w}_j\| \\ \underset{\sim}{z}_{j+1} &= \eta_{j+1}^{-1} \times \underset{\sim}{w}_j \ . \end{aligned}\right\} \tag{4.3}$$

It is well known (cf. [9]) that the eigenvalues of the symmetric tridiagonal matrix

$$J_k = \text{tridiag}\,\{\eta_{j-1}, \xi_j, \eta_j\}$$

are the roots of the polynomial $\tilde{p}_k(\lambda)$ and that the square of the first component of the orthonormalized eigenvectors is the associated weight of the quadrature rule when $\mu_0 = 1$. (This corresponds to a normalization of the weights). The eigenvalues and the first components of each of the normalized eigenvectors can be efficiently computed by the QR method (cf. [5]). It can be shown that the vectors $\{\underset{\sim}{z}_j\}_{j=0}^{k}$ can be obtained directly by orthogonalizing the vectors $\{\underset{\sim}{r}_j\}_{j=0}^{k}$ by an orthogonalization procedure such as the Gram-Schmidt method.

In order to guarantee the numerical stability of the process, the vector $\underset{\sim}{z}_{j+1}$ computed by (4.3) must be re-orthogonalized with respect to all the previously computed $\underset{\sim}{z}_j$'s. This is especially important when it is necessary to know the nodes and weights precisely.

Let

$$\bar{J}_{k+1} = \begin{bmatrix} \xi_1 & \eta_1 & & & & \bigcirc \\ \eta_1 & \xi_2 & \eta_2 & & & \\ & \eta_2 & \cdot & \cdot & & \\ & & \cdot & \cdot & \cdot & \\ & & & \cdot & \cdot & \cdot \\ & \bigcirc & & & \cdot & \cdot & \eta_k \\ & & & & & \eta_k & \bar{\xi}_{k+1} \end{bmatrix}$$

We wish to compute the element $\bar{\xi}_{k+1}$ so that $p_{k+1}(t_0) = 0$, and thus the eigenvalues and eigenvectors of $\bar{J}_{k+1}$ yield the Gauss-Radau rule. Now

$$0 = p_{k+1}(t_0) = (t_0 - \bar{\xi}_{k+1})p_k(t_0) - \eta_k^2 p_{k-1}(t_0) ,$$

and hence

$$\bar{\xi}_{k+1} = t_0 - \eta_k^2 p_{k-1}(t_0)/p_k(t_0) .$$

The quantity $\bar{\xi}_{k+1}$ can be easily computed as follows (cf. [7]):

Solve the equation

$$(J_k - aI)\underset{\sim}{\delta} = \eta_k^2 \underset{\sim}{e}_k$$

where $\underset{\sim}{e}_k^T = (0,0,\ldots,1)$. Then

$$\bar{\xi}_{k+1} = a + \delta_k .$$

It is not necessary to compute the eigenvalues and eigenvectors of $\bar{J}_{k+1}$ to compute upper and lower bounds on μ_s . Let

$$\bar{J}_{k+1} = QTQ^T , \ QQ^T = I_{k+1},$$

where T is the diagonal matrix of eigenvalues of $\bar{J}_{k+1}$ and Q is the matrix of eigenvectors. The vector $Q^T \underset{\sim}{e}_1$, $(\underset{\sim}{e}_1^T = (1,0,\ldots,0))$ consists of the first element of each eigenvector of $\bar{J}_{k+1}$. Hence

$$\mu_s \sim \sum_{i=0}^{k} A_i t_i^s = \underset{\sim}{e}_1^T \bar{J}_{k+1}^s \underset{\sim}{e}_1 .$$

Hence if $s = -2$, then

$$\sum_{i=0}^{k} A_i t_i^{-2} = (\underset{\sim}{e}_1^T \bar{J}_{k+1}^{-1}) (\bar{J}_{k+1}^{-1} \underset{\sim}{e}_1) .$$

Summarizing we have the following algorithm for computing an upper bound on $\|\underset{\sim}{e}\|_2$.

1) Compute $\underset{\sim}{r}_0 = \underset{\sim}{b} - A\,\underset{\sim}{\xi}$

2) Apply the Lanczos procedure to A with $\underset{\sim}{z}_0 = (\|\underset{\sim}{r}_0\|)^{-1} \times \underset{\sim}{r}_0$ (see (4.3)).

3) For given a, $0 < a \leq \lambda_1$, determine $\bar{\xi}_{k+1}$ to construct the matrix $\bar{J}_{k+1}$.

4) Solve $\bar{J}_{k+1}\,\underset{\sim}{g} = \underset{\sim}{e}_1$,

5) Then $\mu_{-2} \leq \mu_0 \times \underset{\sim}{g}^T \underset{\sim}{g}$.

In order to compute a lower bound for μ_{-2}, one computes the element $\bar{\xi}_{k+1}$ so that $b(\geq \lambda_n)$ is an eigenvalue of $\bar{J}_{k+1}$.

Of course, it is most desirable if it is possible to use the information in the vectors $\{\underset{\sim}{z}_j\}_{j=0}^{k}$ for improving the approximate solution $\underset{\sim}{\xi}$. In section 5, we show how this can be done by relating the conjugate gradient method of Hestenes and Stiefel to the Lanczos process and then modifying the conjugate gradient method.

5. IMPROVING THE APPROXIMATE SOLUTION.

In the previous section, we showed how bounds can be obtained for $\|\underset{\sim}{e}\|$. In obtaining the bounds, it was necessary to compute $\{\underset{\sim}{r}_i\}_{i=0}^{k}$ or equivalently $\{\underset{\sim}{z}_i\}_{i=0}^{k}$. We now consider the problem of determining an improved estimate of the solution vector $\underset{\sim}{x}$ and determining error bounds for the improved estimate.

Let us denote the improved estimate as $\tilde{\underset{\sim}{x}}$ so that

$$\begin{aligned} \tilde{\underset{\sim}{x}} &= \underset{\sim}{\xi} + \sum_{i=0}^{k} \beta_i\,\underset{\sim}{r}_i \\ &= \underset{\sim}{\xi} + \sum_{i=0}^{k} \beta_i\,A^i \underset{\sim}{r}_0 \qquad (5.1) \\ &\equiv \underset{\sim}{\xi} + P_k(A)\underset{\sim}{r}_0 \end{aligned}$$

where $P_k(A)$ denotes a polynomial of degree k in the matrix A. Define

$$\Psi(\underset{\sim}{\beta}) \equiv (\tilde{\underset{\sim}{x}} - \underset{\sim}{x})^T A\,(\tilde{\underset{\sim}{x}} - \underset{\sim}{x}) \qquad (5.2)$$

We consider the following problems.

Problem 2 A.

Determine $\hat{\underset{\sim}{\beta}}$ so $\Psi(\hat{\underset{\sim}{\beta}}) = \min$.

Problem 2 B.

Determine upper and lower bounds for $\Psi(\hat{\underset{\sim}{\beta}})$.

If $\underset{\sim}{e}_{k+1} = \tilde{\underset{\sim}{x}} - \underset{\sim}{x}$, then

$$\Psi(\hat{\underset{\sim}{\beta}}) = \underset{\sim}{e}_{k+1}^T A\,\underset{\sim}{e}_{k+1}.$$

The solution to Problem 2 A is well known (cf. [10]). The construction of the vector $\tilde{\underset{\sim}{x}}$ is accomplished via the conjugate gradient method of Hestenes and Stiefel. There are many alternative forms of this algorithm, we give one which is quite easy to implement. Then the vector which is denoted by $\underset{\sim}{x}_{k+1}$ is the vector $\tilde{\underset{\sim}{x}}$ which we have sought.

<u>Algorithm.</u>

Let $\underset{\sim}{x}_0 = \underset{\sim}{\xi}$ be a given vector and arbitrarily define $\underset{\sim}{p}_{-1}$. For $j = 0,1,\ldots,k$

(1) Compute $\rho_j = \underset{\sim}{b} - A\underset{\sim}{x}_0$

(2) Compute
$$b_j = \frac{\underset{\sim}{\rho}_j^T \underset{\sim}{\rho}_j}{\underset{\sim}{\rho}_{j-1}^T \underset{\sim}{\rho}_{j-1}} \quad , \quad j \geq 1 \quad , \quad (b_0 = 0)$$

$$\underset{\sim}{p}_j = \underset{\sim}{\rho}_j + b_j \underset{\sim}{p}_{j-1} \; .$$

(3) Compute
$$a_j = \frac{\underset{\sim}{\rho}_j^T \underset{\sim}{\rho}_j}{\underset{\sim}{p}_j^T A \underset{\sim}{p}_j}$$

$$\underset{\sim}{x}_{j+1} = \underset{\sim}{x}_j + a_j \underset{\sim}{p}_j \; .$$

It is not necessary to re-compute the residual vector $\underset{\sim}{\rho}_j$ at each stage, it is often advantageous to compute it recursively from

$$\underset{\sim}{\rho}_{j+1} = \underset{\sim}{\rho}_j - a_j A \underset{\sim}{p}_j \; .$$

Note then that the vectors $\{\underset{\sim}{r}_j\}_{j=0}^k$ are never explicitly computed. Furthermore, using the conjugate gradient method it is an easy matter to compute $\underset{\sim}{x}_{k+2}$ from $\underset{\sim}{x}_{k+1}$ and auxiliary vectors.

A short manipulation shows that

$$A\underset{\sim}{\rho}_j = -\frac{b_j}{a_{j-1}} \underset{\sim}{\rho}_{j-1} + \left(\frac{1}{a_j} + \frac{b_j}{a_{j-1}}\right) \underset{\sim}{\rho}_j - \frac{1}{a_j} \underset{\sim}{\rho}_{j+1} \; . \quad (j = 0,1,..,k)$$

Thus

$$A[\underset{\sim}{\rho}_0, \underset{\sim}{\rho}_1, \ldots, \underset{\sim}{\rho}_k] \equiv [\underset{\sim}{\rho}_0, \underset{\sim}{\rho}_1, \ldots, \rho_{k+1}] \, K_k$$

where K is a tri-diagonal matrix. Since $\underset{\sim}{\rho}_0 = \underset{\sim}{r}_0$, it can be shown that K_n is similar to the tri-diagonal matrix J_n. After some manipulations then, we have

$$\left.\begin{aligned} \xi_{j+1} &= \left(\frac{1}{a_j} + \frac{b_j}{a_{j-1}}\right) && (j = 0,1,..,k-1) \\ \eta_j^2 &= \frac{b_j}{a_{j-1}^2} && (j = 1,..,k-1) \end{aligned}\right\} \qquad (5.3)$$

Using (5.1), we have

$$(\tilde{\underset{\sim}{x}} - \underset{\sim}{x}) = (\underset{\sim}{\xi} - \underset{\sim}{x}) + P_k(A)\underset{\sim}{r}_0$$

$$= -A\underset{\sim}{r}_0 + P_k(A)\underset{\sim}{r}_0 \; .$$

Since

$P_k(A)\underset{\sim}{r}_0 = \sum_{i=0}^{k} \beta_i \underset{\sim}{r}_i$, a short calculation shows

$$\Psi(\underset{\sim}{\beta}) = \mu_{-1} - 2\underset{\sim}{\beta}^T \mu + \underset{\sim}{\beta}^T M_1 \underset{\sim}{\beta} \qquad (5.4)$$

where

$$\mu_j = \underset{\sim}{r}_0^T A^j \underset{\sim}{r}_0 \qquad (j = -1,0,\ldots,2k+1)$$

$$\underset{\sim}{\mu}^T = (\mu_0, \mu_1, \ldots, \mu_k)$$

$$\underset{\sim}{\beta}^T = (\beta_0, \beta_1, \ldots, \beta_k)$$

and

$$M_1 = \begin{bmatrix} \mu_1, \mu_2, & \cdot & \cdot & \cdot & , \mu_{k+1} \\ \mu_2, \mu_3, & \cdot & \cdot & \cdot & , \mu_{k+2} \\ \cdot & & & & \\ \cdot & & & & \\ \cdot & & & & \\ \mu_{k+1}, \mu_{k+2} & \cdot & \cdot & \cdot & , \mu_{2k+1} \end{bmatrix} . \tag{5.5}$$

The vector which minimizes $\Psi(\underset{\sim}{\beta})$ is denoted as $\hat{\underset{\sim}{\beta}}$ and satisfies the equation

$$M_1 \hat{\underset{\sim}{\beta}} = \underset{\sim}{\mu} \tag{5.6}$$

and

$$\Psi(\hat{\underset{\sim}{\beta}}) = \mu_{-1} - \underset{\sim}{\mu}^T M_1^{-1} \underset{\sim}{\mu} \tag{5.7}$$

It is not necessary to explicitly compute $\hat{\underset{\sim}{\beta}}$, since it is implicitly computed by the conjugate gradient method. The bounds for $\Psi(\hat{\underset{\sim}{\beta}})$ are determined by evaluating the quadratic form $\underset{\sim}{\mu}^T M_1^{-1} \underset{\sim}{\mu}$ and by determining upper and lower bounds for μ_{-1}. We shall now consider this problem.

By (3.1), we see that if $\varphi(\lambda) = \lambda^{-1}$ and $m = 1$, the remainder term yields

$$R[\lambda^{-1}] = - \eta^{-(2k+2)} \int_a^b (\lambda - z_1) \left[\prod_{i=1}^{k} (\lambda - t_i) \right]^2 d\,\alpha(\lambda) .$$

Hence if $z_1 = a > 0$, the Gauss-Radau type quadrature rule yields an upper bound for $\mu_{-1} = \int_a^b \lambda^{-1} d\,\alpha(\lambda)$. Again it is possible to use the Lanczos process and/or conjugate gradient method for determining the upper and lower bounds for μ_{-1}.

Let $\bar{J}_{k+1}$ be the $(k+1)\times(k+1)$ matrix whose eigenvalues are $t_0 = a$ and $\{t_i\}_{i+1}^k$. Then

$$\bar{J}_{k+1} = Q \Lambda Q^T .$$

As noted,

$$\mu_{-1} \le \sum_{i=0}^{k} \frac{A_i}{t_i} = \sum_{i=1}^{k+1} \frac{q_{1j}^2}{\lambda_i} \tag{5.8}$$

where q_{1j} is the first element of each eigenvector of $\bar{J}_{k+1}$. Therefore, by solving

$$\bar{J}_{k+1} \underset{\sim}{g} = \underset{\sim}{e}_1$$

and then computing $\underset{\sim}{e}_1^T \underset{\sim}{g} = g_1$, we have an upper bound for μ_{-1}. That is

$$\mu_{-1} \leq \mu_0 \, g_1 = \|\underset{\sim}{r}_0\|^2 g_1 \tag{5.9}$$

By changing the (k+1),(k+1) element of $\bar{J}_{k+1}$ so that b is an eigenvalue of $\bar{J}_{k+1}$ we can determine a lower bound for μ_{-1}.

Finally, we shall show that

$$\underset{\sim}{\mu}^T M_1^{-1} \underset{\sim}{\mu} = \mu_0 \{J_{k+1}^{-1}\}_{11} = \|\underset{\sim}{r}_0\|^2 \{J_{k+1}^{-1}\}_{11} \tag{5.10}$$

Let us denote

$$M_1 = \begin{bmatrix} \mu_1, & \mu_2, & \cdot & \cdot & \cdot & , \mu_{k+1} \\ \mu_2, & \mu_3, & \cdot & \cdot & \cdot & , \mu_{k+2} \\ \cdot & & & & & \\ \cdot & & & & & \\ \cdot & & & & & \\ \mu_{k+1}, & \cdot & \cdot & \cdot & \cdot & , \mu_{2k+1} \end{bmatrix} \tag{5.11}$$

Recall that $\{\tilde{p}_j(\lambda)\}$ indicates the orthonormal polynomials which satisfy the relationship

$$\eta_{j+1}\tilde{p}_{j+1}(\lambda) = (\lambda - \xi_{j+1})\tilde{p}_j(\lambda) - \eta_j \tilde{p}_{j-1}(\lambda) \tag{5.12}$$

We write

$$\tilde{p}(\lambda) = (\tilde{p}_0(\lambda), \tilde{p}_1(\lambda), \ldots, \tilde{p}_k(\lambda))$$

and hence

$$\underset{\sim}{\tilde{p}}(\lambda) = \Phi \begin{bmatrix} 1 \\ \lambda \\ \lambda^2 \\ \cdot \\ \cdot \\ \cdot \\ \lambda^k \end{bmatrix} \equiv \Phi \, \underset{\sim}{\omega}(\lambda)$$

where Φ is a $(k+1) \times (k+1)$ lower triangular matrix, and

$$\omega(t)^T = (1, \lambda, \lambda^2, \ldots, \lambda^k) .$$

Since the polynomials are orthonormal,

$$\int \underset{\sim}{\tilde{p}}(\lambda) \, \underset{\sim}{p}(\lambda)^T d\,\alpha(\lambda) = I_{k+1} .$$

Then

$$I_{k+1} = \int \underset{\sim}{\tilde{p}}(\lambda) \, \underset{\sim}{p}(\lambda)^T d\,\alpha(\lambda) = \int \Phi \, \underset{\sim}{\omega}(\lambda)(\underset{\sim}{\omega}(\lambda))^T \Phi^T d\,\alpha(\lambda)$$

$$= \Phi \, M_0 \, \Phi^T .$$

Hence

$$M_0 = \Phi^{-1}(\Phi^T)^{-1} . \tag{5.13}$$

Now note that

$$\Phi M_1 \Phi^T = \Phi \int \lambda \, \underset{\sim}{\omega}(\lambda)(\underset{\sim}{\omega}(\lambda))^T d\,\alpha(\lambda)\, \Phi^T$$

$$= \int \lambda \, \tilde{\underset{\sim}{p}}(\lambda)(\tilde{\underset{\sim}{p}}(\lambda))^T \, d\,\alpha(\lambda) \quad . \tag{5.14}$$

Using the relationship (5.12), we have

$$\lambda \, \tilde{p}_j(\lambda) = \eta_{j+1} \, \tilde{p}_{j+1}(\lambda) + \xi_{j+1} \, \tilde{p}_j(\lambda) + \eta_j \, \tilde{p}_{j-1}(\lambda)$$

and this in combination with (5.14) leads to

$$\Phi M_1 \Phi^T = J_{k+1} = \begin{bmatrix} \xi_1 & \eta_1 & & & \bigcirc \\ \eta_1 & \cdot & \cdot & & \\ & \cdot & \cdot & \cdot & \\ & & \cdot & \cdot & \eta_k \\ \bigcirc & & & \eta_k & \xi_{k+1} \end{bmatrix} . \tag{5.15}$$

Thus

$$M_1^{-1} = \Phi^T J_{k+1}^{-1} \Phi \tag{5.16}$$

and hence by (5.13) and (5.16)

$$M_0 M_1^{-1} M_0 = \Phi^{-1} J_{k+1}^{-1} (\Phi^T)^{-1} \tag{5.17}$$

Now

$$\{M_0 M_1^{-1} M_0\}_{11} = \underset{\sim}{\mu}^T M_1^{-1} \underset{\sim}{\mu} ,$$

and Φ is a lower triangular matrix with $\{\Phi\}_{11} = \frac{1}{\sqrt{\mu_0}}$. Thus,

$$\underset{\sim}{\mu}^T M_1^{-1} \underset{\sim}{\mu} = \mu_0 \{J_{k+1}^{-1}\}_{11} = \|\underset{\sim}{r}_0\|^2 \{J_{k+1}^{-1}\}_{11} .$$

As noted earlier, the matrix J_{k+1} can be computed explicitly when using the conjugate gradient method.

Summarizing then, we have the following.

Using the conjugate gradient method we are able to estimate the error on the approximation $\underset{\sim}{\xi}$, we can improve the approximation by the conjugate gradient method and this yields an approximation $\underset{\sim}{x}_{k+1}$, and finally we are able to obtain upper and lower bounds on $\underset{\sim}{e}_{k+1}^T A \, \underset{\sim}{e}_{k+1}$.

6. COMPUTATIONAL RESULTS.

In order to evaluate the effectiveness of these error bounds, the algorithms were tried with

$$A = \begin{bmatrix} 2 & -1 & & & & \\ -1 & 2 & -1 & & \bigcirc & \\ & -1 & 2 & & & \\ & & \cdot & & & \\ & & & \cdot & & \\ & \bigcirc & & & \cdot & \\ & & & & 2 & -1 \\ & & & & -1 & 2 \end{bmatrix}_{n \times n}$$

$$\underset{\sim}{b}^T = [0,0,\ldots,0]$$

$$\underset{\sim}{\tilde{x}}^T = [\xi_1,\ldots,\xi_n] \qquad \xi_i \sim U(-1,1)$$

for $n = 25,50,100$. At each stage k of the conjugate gradient algorithm, the upper and lower estimates of $\underset{\sim}{e}_k^T A \underset{\sim}{e}_k$ were computed as well as the actual value of $\underset{\sim}{e}_k^T A \underset{\sim}{e}_k$. In addition, the bounds on $\|\underset{\sim}{e}_0\|_2$ were computed. The results of these computations are graphed below using logarithmic scaling for the norms of the errors. The conjugate gradient algorithm was used both with and without re-orthogonalization, but there did not seem to be any noticeable difference between the two computations.

From the graphs, it can be seen that although the value of $\underset{\sim}{e}_k^T A \underset{\sim}{e}_k$ decreases monotonically, the estimates for the A-norm of the error do not. The other notable observation is that the upper estimate of $\underset{\sim}{e}_k^T A \underset{\sim}{e}_k$ seems to be much better than the lower estimate. We have no explanation for these phenomena.

These computations were performed using double precision under the FORTG compiler on an IBM 370/168 machine.

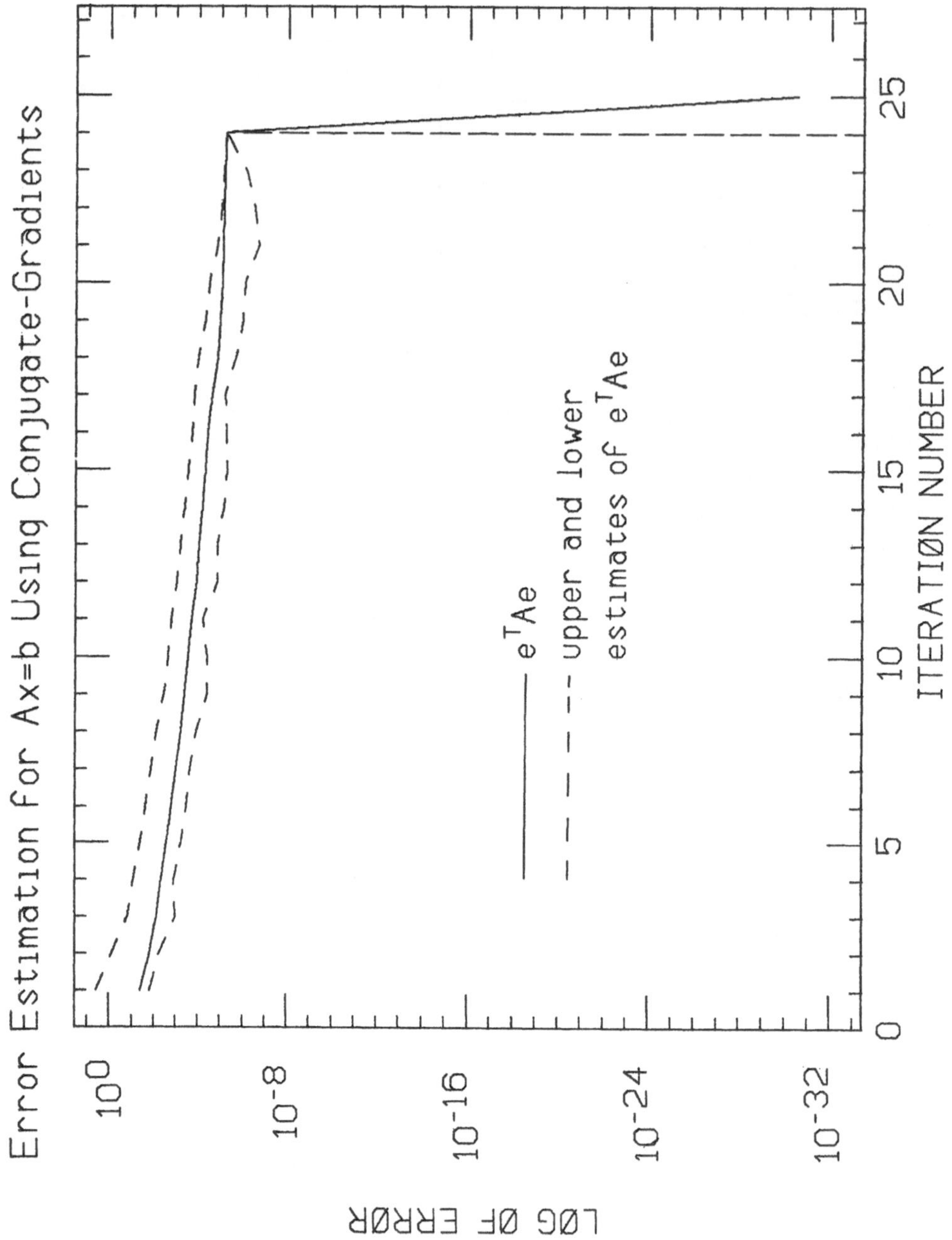

Estimates of e^TAe
(without re-orthogonalization)

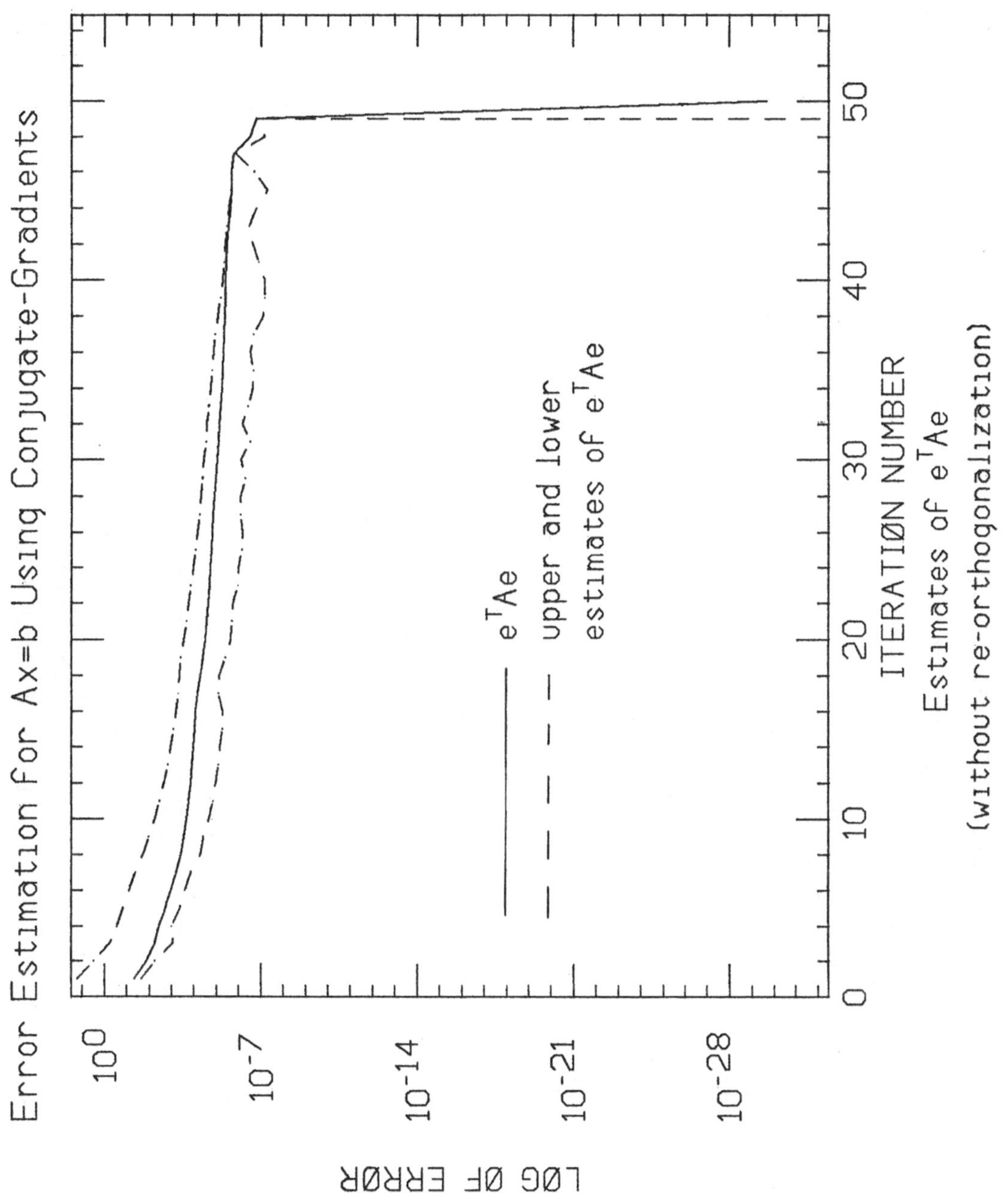
Error Estimation for Ax=b Using Conjugate-Gradients
LØG ØF ERRØR
10^{0}
10^{-7}
10^{-14}
10^{-21}
10^{-28}
0
10
20
30
40
50
$e^{T}Ae$
upper and lower
estimates of $e^{T}Ae$
ITERATIØN NUMBER
Estimates of $e^{T}Ae$
(without re-orthogonalization)

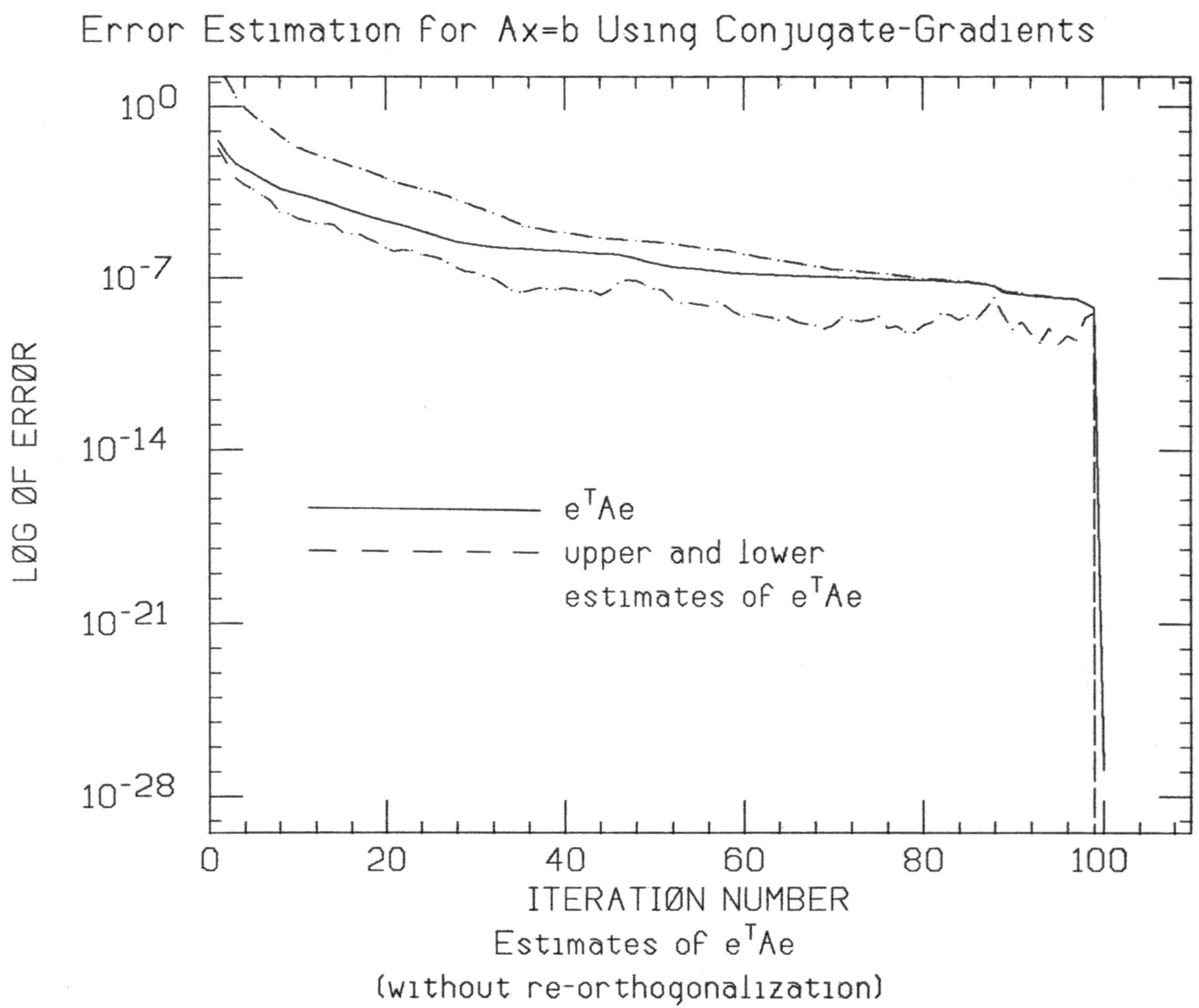
Error Estimation for Ax=b Using Conjugate-Gradients
LØG ØF ERRØR
10^0
10^-7
10^-14
10^-21
10^-28
0
20
40
60
80
100
e^T Ae
upper and lower
estimates of e^T Ae
ITERATIØN NUMBER
Estimates of e^T Ae
(without re-orthogonalization)

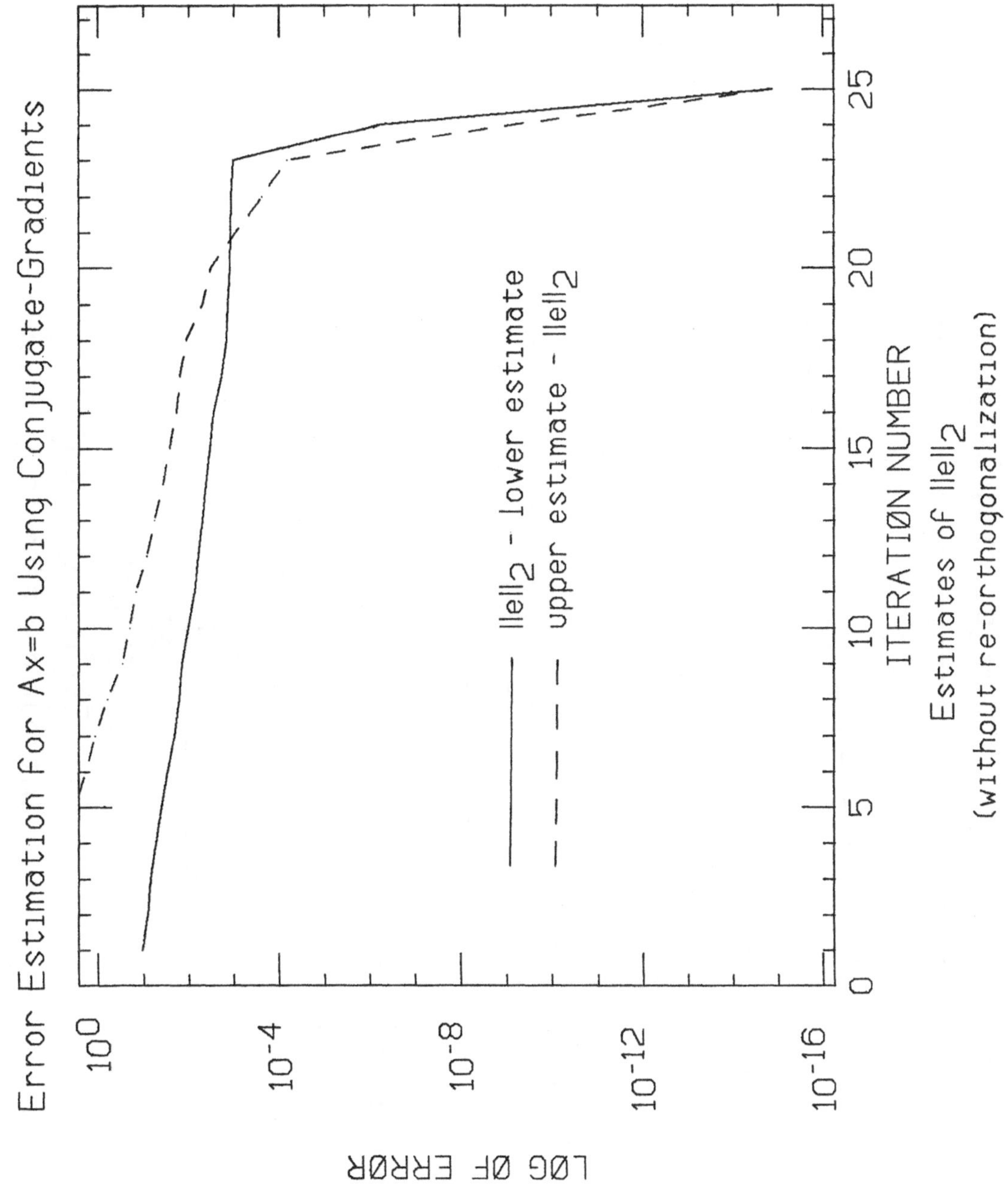
Error Estimation for Ax=b Using Conjugate-Gradients
LOG OF ERROR
10^{0}
10^{-4}
10^{-8}
10^{-12}
10^{-16}
0
5
10
15
20
25
ITERATION NUMBER
$\|e\|_2$ - lower estimate
upper estimate - $\|e\|_2$
Estimates of $\|e\|_2$
(without re-orthogonalization)

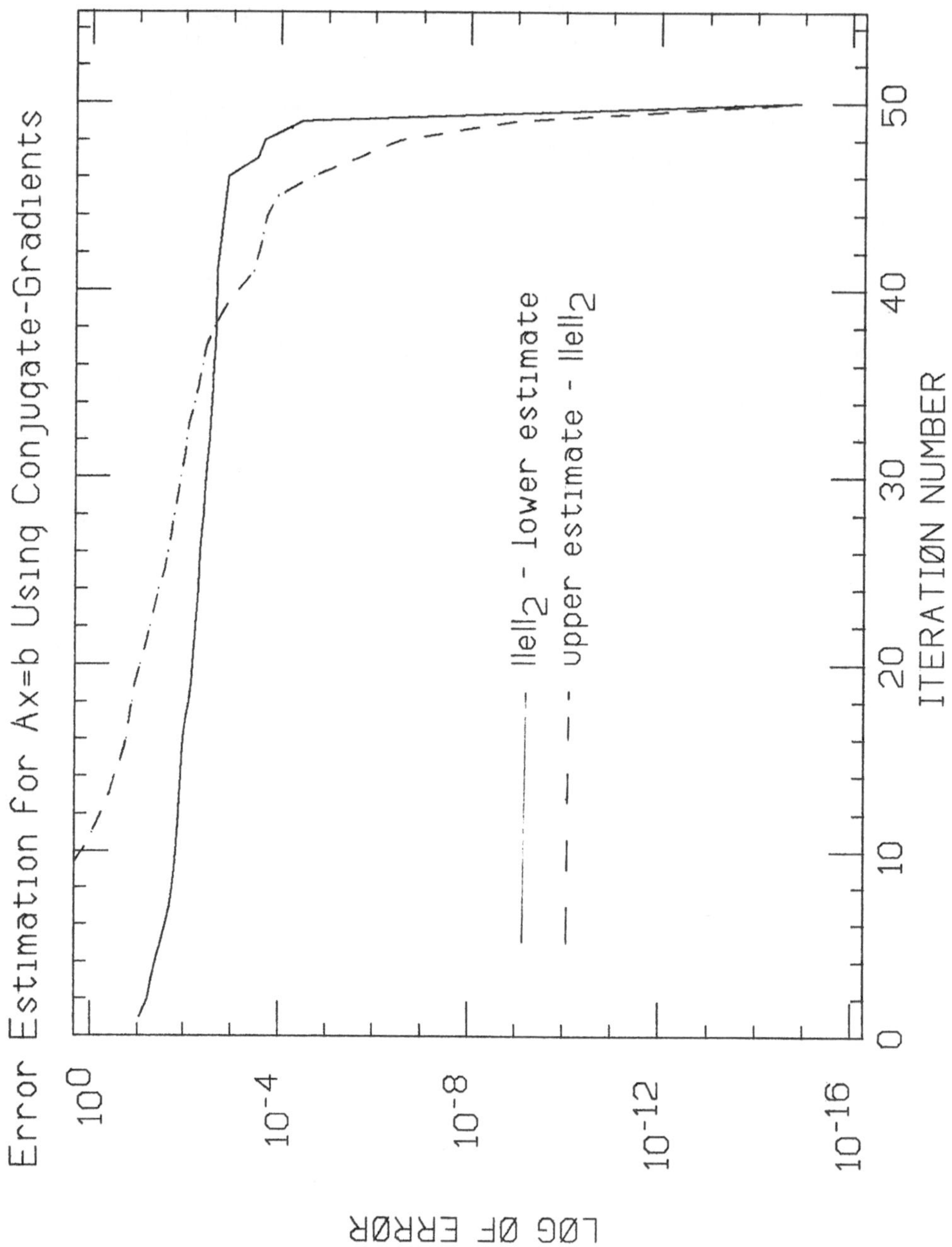

Estimates of $\|e\|_2$
(without re-orthogonalization)

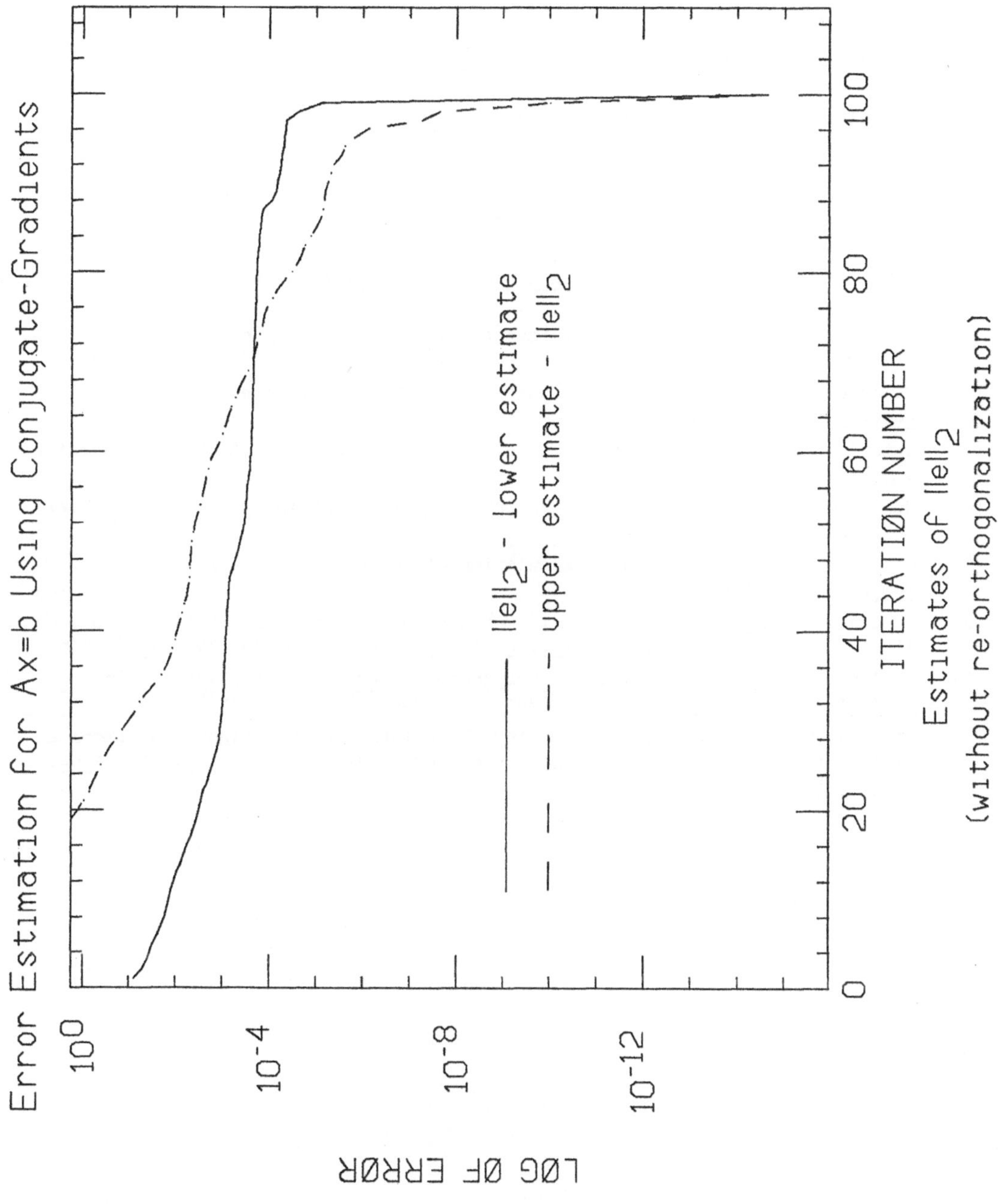
Error Estimation for Ax=b Using Conjugate-Gradients
10^{0}
10^{-4}
10^{-8}
10^{-12}
LOG OF ERROR
0
20
40
60
80
100
ITERATION NUMBER
$\|e\|_2$ - lower estimate
upper estimate - $\|e\|_2$
Estimates of $\|e\|_2$
(without re-orthogonalization)

References:

[1] Aheizer, N.I. and Krein, M., Some Questions in the Theory of Moments, Amer. Math. Soc., Providence, Rhode Island, 1962.

[2] Bartels, R. and Golub, G.H., "Stable numerical methods for obtaining the Chebyshev solution to an overdetermined system of equations, Comm.A.C.M. 11 (1968), 401-406.

[3] Dahlquist, G., Eisenstat, S., and Golub, G.H., "Bounds for the error of linear systems of equations using the theory of moments," J. Math. Anal. Appl. 37 (1972), 151-166.

[4] Dantzig, G.B., Linear Programming and Extensions, Princeton Univ. Press, Princeton, N.J., 1963.

[5] Francis, J.G.G., "The QR transformation: a unitary analogue to the LR transformation. I,II," Computer J. 4 (1961), 265-271; 322-345.

[6] Gautschi, W., "Construction of Gauss-Christoffel quadrature formulas," Math. Comp. 22 (1968), 251-270.

[7] Golub, G.H., "Some modified matrix eigenvalue problems," SIAM Rev. 15 (1973), 318-334.

[8] Golub, G.H., "Bounds for matrix moments," Rocky Mountain J. of Math. 4 (1974), 207-211.

[9] Golub, G.H. and Welsch, J., "Calculation of Gauss quadrature rules," Math. Comp. (1969), 221-230.

[10] Hestenes, M.R. and Stiefel, E.L., "Methods on conjugate gradients for solving linear systems," J. Res. Nat. Bur. Standards, Section B 49 (1952), 409-436.

[11] Karlin, S. and Studden, W.J., "Tchebysheff Systems: With Application in Analysis and Statistics, Interscience Publishers, New York, 1966.

[12] Lanczos, C., "An iteration method for the solution of the eigenvalue problem of linear differential and integral operators," J. Res. Nat. Bur. Standards, Section B 45 (1950), 255-282.

One Sided L_1-Approximation as a Problem of Semi-Infinite Linear Programming

W.Krabs, Darmstadt

1. Introduction: General Remarks on One Sided L_1-Approximation

L_1-approximation of functions, say, by polynomials or other types of simple functions does not occur as frequently as, for instance, Chebychev approximation. Occasionally, however, L_1-approximation, in particular from one side, arises in a natural way from a given problem. So in [5] Marsaglia (see also [3] for a short description of the problem) deals with the approximation of the density of a probability distribution from below by linear combinations of given density functions such that the area between the corresponding curves becomes as small as possible. This is a typical problem of one sided L_1-approximation which amounts to maximizing a linear form on the nonnegative orthant of R^n subject to infinitely many linear constraints. Marsaglia replaces this problem with a problem of linear programming by only requiring a finite number of the constraints. For the solution of this approximating problem a condensed form of the revised simplex method is used.

Best one sided L_1-approximation of functions on finite intervals by polynomials has been investigated by Bojanic and DeVore in [2] without referring to optimization. They prove existence of best approximations, if the function to be approximated is Lebesgue integrable and bounded from below. They further show by a counterexample that continuity of the function to be approximated is not enough to guarantee uniqueness of best approximations. Uniqueness holds, if in addition to continuity on the closed interval differentiability in the interior is required. By relating the problem of one sided L_1-approximation to best generalized Gauß quadrature rules (by Markov, Radau and Lobatto) it is shown to be reducible to Hermite interpolation with known nodes, if the function to be approximated is continuous and has an n-th derivative of one sign in the interior of the interval. Several more results of this type are given also including approximation by trigonometric polynomials.

Recently semi-infinite linear programming problems in Haar spaces have been investigated in [6] and solved by a method of moments due to Markov. The results include as a special case, the above statement by Bojanic and DeVore concerning one sided L_1-approximation of continuous functions having an n-th derivative of one sign.

In this note we consider a special problem of one sided L_1-approximation which arises with the approximate solution of weakly singular integral equations(see[1]).

The problem is stated as a semi-infinite linear programming problem such that the corresponding dual problem satisfies the generalized Slater condition. This ensures the solvability of the problem and the equality of the extreme values of the two problems. The solvability of the dual problem does not seem to be guaranteed. By requiring the function to be approximated to be differentiable at the right end of the interval (which is not very restrictive in the view of applications) the L_1-approximation problem can be replaced by a semi-infinite linear programming problem which satisfies the generalized Slater condition. So the dual problem takes on the well known form and becomes solvable. The solution of the problem can be reduced to interpolation, however, with nodes whose number and location are not known in advance.

2. On a Special One Sided L_1-Approximation Problem

2.1. Statement of the Problem and Solvability

In connection with the approximate solution of weakly singular integral equations by a double approximation (see[1]) the following one sided L_1-approximation problem arises:
Given a number $b\in(o,1)$ and a decreasing function $g\in L^1[o,b]\cap C(o,b]$, we are looking for a polynomial of degree $m\geq o$,

$$p_m(x,r)=\sum_{k=o}^{m}(b-r)^k x_k,\quad r\in[o,b],$$

$$x=(x_o,\dots,x_m)\in R^{m+1}, \tag{2.1}$$

which is decreasing and satisfies

$$p_m(x,r)\leq g(r),\quad r\in[o,b], \tag{2.2a}$$

$$p_m(x,b)=g(b) \tag{2.2b}$$

such that its L_1-deviation from g,

$$\| g-p_m(x,\cdot) \|_1 = \int_o^b g(r)-p_m(x,r)dr,$$

becomes as small as possible.
We assume $x_k \geq o$ for all $k=1,\ldots,m$ which ensures that $p_m(x,\cdot)$ given by (2.1) is decreasing. Condition (2.2b) is equivalent to $x_o=g(b)$. Hence the problem can be rewritten as

Problem (P): Maximize

$$\int_o^b p_m(x,r)dr-bg(b) = \sum_{k=1}^{m} \frac{b^{k+1}}{k+1} x_k$$

under the restrictions

$$\sum_{k=1}^{m} (b-r)^k x_k \leq g(r)-g(b),\ r\in[o,b], \tag{2.3a}$$

$$x_k \geq o \text{ for } k=1,\ldots,m. \tag{2.3b}$$

We consider this problem at first as an infinite linear programming problem (see, for instance,[4]). Then it can be assigned a dual problem in the form of

Problem (D): Minimize

$$\int_o^b (g(t)-g(b))y(t)dt$$

under the conditions

$$y\in L^\infty[o,b],\ y\geq o \text{ almost everywhere on } [o,b], \tag{2.4a}$$

$$\int_o^b (b-t)^k y(t)dt \geq \frac{b^{k+1}}{k+1},\ k=1,\ldots,m. \tag{2.4b}$$

If we choose $x_k=o$ for $k=1,\ldots,m$, then the restrictions (2.3) are satisfied i.e., the Problem (P) is feasible. If for any $\varepsilon>o$ we put $y=1+\varepsilon$, then y satisfies (2.4a) and (2.4b) is even satisfied with strict inequalities for all k, i.e., the Problem (D) fulfills the generalized Slater condition. Hence, by a well-known result of the theory of infinite linear programming (see, for instance,[4]), the Problem (P) is solvable and the extreme values of both problems coincide, i.e., we have that

$$\gamma := \sup\{ \sum_{k=1}^{m} \frac{b^{k+1}}{k+1} x_k \mid x\in R^m \text{ satisfies } (2.3a+b)\} \tag{2.5}$$

is assumed for some $x\in R^m$ that meets (2.3a+b) and that

$$\gamma = \inf\{ \int_o^b (g(t)-g(b))y(t)dt \mid y \text{ satisfies } (2.4a+b)\}. \tag{2.6}$$

Furthermore, we obtain the (obvious) inclusion statement

$$o \le \gamma \le \int_o^b g(t)\,dt - bg(b). \tag{2.7}$$

2.2 Reduction to a Problem of Semi-Infinite Programming

Let $x \in R^m$ be feasible for Problem (P). Then we put

$$q_m(x,r) = g(b) + \sum_{k=1}^{m} (b-r)^k x_k$$

and obtain

$$q_m(x,o) \le g(b) + \frac{m+1}{b} \sum_{k=1}^{m} \frac{b^{k+1}}{k+1} x_k$$

which implies

$$q_m(x,o) \le g(b) + \frac{m+1}{b}\gamma,$$

hence,

$$\sum_{k=1}^{m} (b-r)^k x_k \le q_m(x,o) - g(b) \le \frac{m+1}{b}\gamma, \in [o,b].$$

Let $\alpha = \lim_{r \to o+} g(r)$. Then there is a greatest $\varepsilon \in [o,b]$ such that

$$g(\varepsilon) = \min(\alpha, \frac{m+1}{b}\gamma).$$

If we define

$$g_\varepsilon(r) = \begin{cases} g(\varepsilon), & r \in [o,\varepsilon], \\ g(r), & r \in [\varepsilon,b], \end{cases}$$

then Problem (P) becomes equivalent to

Problem (P_ε): Maximize

$$\sum_{k=1}^{m} \frac{b^{k+1}}{k+1} x_k$$

subject to

$$\sum_{k=1}^{m} (b-r)^k x_k \le g_\varepsilon(r) - g(b), \quad r \in [o,b] \tag{2.3a)$_\varepsilon$}$$

$$x_k \ge o, \quad k=1,\dots,m. \tag{2.3b}$$

This is a typical semi-infinite linear programming problem in $C[o,b]$ instead of $L^1[o,b]$ which is solvable and does not have a duality gap. In order to ensure the solvability of the corresponding dual problem it would suffice to satisfy $(2.3a)_\varepsilon$ as strict inequality for all

$r\epsilon[o,b]$ which is impossible for $r=o$. This little difficulty can be overcome by assuming that g be strictly decreasing and differentiable at b which is no serious restriction in the view of applications. In this case the restrictions $(2.3a)_\varepsilon$ are equivalent to

$$\sum_{k=1}^{m} (b-r)^{k-1} x_k \le c_\varepsilon(r) = \begin{cases} \dfrac{g_\varepsilon(r)-g(b)}{b-r}, & r\epsilon[o,b), \\ -g'(b) \text{ for } r=b. \end{cases} \qquad (2.8)_\varepsilon$$

The function $c_\varepsilon = c_\varepsilon(r)$ is continuous and positive on $[o,b]$ so that the restrictions $(2.8)_\varepsilon$ are strictly satisfied for all $r\epsilon[o,b]$, if $x=\Theta_m$. This implies the solvability of the dual problem in the following form (see[4])

<u>Problem (D_ε):</u> Minimize

$$\sum_{i=1}^{m} c_\varepsilon(r_i) y_i \quad (c_\varepsilon(r) \text{ given by } (2.8)_\varepsilon)$$

subject to

$$r_i \epsilon [o,b], y_i \ge o \text{ for } i=1,\dots,m \qquad (2.9a)$$

and

$$\sum_{i=1}^{m} (b-r_i)^{k-1} y_i \ge \frac{b^{k+1}}{k+1}, \; k=1,\dots,m. \qquad (2.9b)$$

In addition to the solvability and strong duality relation of the Problems (P_ε) and (D_ε) we have the

<u>Theorem:</u> Let $g\epsilon L^1[o,b]\cap C(o,b]$ be strictly decreasing and differentiable at b and let $\hat{x}$ with $(2.3a)_\varepsilon$, (2.3b) and $(\hat{r}_i,\hat{y}_i)_{i=1,\dots,m}$ with (2.9a), (2.9b) be given. Then the following two statements are equivalent:

a) $\hat{x}$ and $(\hat{r}_i,\hat{y}_i)_{i=1,\dots,m}$ solve Problem (P_ε) and (D_ε), respectively.

b) The two implications

$$\hat{x}_k > o \;\rightarrow\; \sum_{i=1}^{m} (b-\hat{r}_i)^{k-1} \hat{y}_i = \frac{b^{k+1}}{k+1} \qquad (2.10a)$$

and

$$\hat{y}_i > o \;\rightarrow\; \sum_{k=1}^{m} (b-\hat{r}_i)^{k-1} x_k = c_\varepsilon(\hat{r}_i) \qquad (2.10b)$$

hold.

Finally we consider the special case m=1. Then $(2.8)_\varepsilon$ reads

$$x_1 \le c_\varepsilon(r) = \begin{cases} \dfrac{g_\varepsilon(r)-g(b)}{b-r} & \text{for } r\epsilon[o,b), \\ -g'(b) & \text{for } r=b \end{cases}$$

and Problem (P_ε) consists of finding a maximal $x_1 \ge o$ that satisfies these inequalities. The solution is given by

$$\hat{x}_1 = \min_{r\epsilon[o,b]} c_\varepsilon(r).$$

The dual Problem (D_ε) consists of minimizing $c_\varepsilon(r_1)y_1$ for $y_1 \ge \frac{b^2}{2}$ and some $r_1 \epsilon [o,b]$. A solution is given by $(\frac{b^2}{2}, r_1)$ with $c_\varepsilon(r_1) = \min_{r\epsilon[o,b]} c_\varepsilon(r)$ and the common extreme value is given by

$$\gamma = \frac{b^2}{2} \min_{r\epsilon[o,b]} c_\varepsilon(r) = \frac{b^2}{2}\hat{x}_1.$$

If in addition g is convex, then

$$-g'(b) = \min_{r\epsilon[o,b]} c_\varepsilon(r)$$

and

$$\gamma = -\frac{b^2}{2} g'(b).$$

3. References.

[1] Anselone,Ph.M. and W.Krabs: Approximate Solution of Weakly Singular Integral Equations. Preprint No.418,TH Darmstadt, Mai 1978. To appear in: Journal of Integral Equations.

[2] Bojanic,R. and R.DeVore: On Polynomials of Best One Sided Approximation. L'Enseignement Mathematique,Ser.2,tome 12,1966,139-164.

[3] Collatz,L. und W.Krabs: Approximationstheorie. Teubner-Verlag Stuttgart 1973,S.32/33.

[4] Krabs,W.: Optimierung und Approximation. Teubner-Verlag: Stuttgart 1975.

[5] Marsaglia,G.: One Sided Approximations by Linear Combinations of Functions. In: Approximation Theory, edited by A.Talbot, New York-London 197o, 233-242.

[6] Sturm,N.: Semi-infinite Optimierung. Fachverlag für Revisions- und Treuhandfragen: Schwarzenbek 1978.

Lecture Notes in Economics and Mathematical Systems

For information about Vols. 1–104 please contact your bookseller or Springer-Verlag

Vol. 105: Optimal Control Theory and its Applications. Part I. Proceedings 1973. Edited by B. J. Kirby. VI, 425 pages. 1974.

Vol. 106: Optimal Control Theory and its Applications. Part II. Proceedings 1973. Edited by B. J. Kirby. VI, 403 pages. 1974.

Vol. 107: Control Theory, Numerical Methods and Computer Systems Modeling. International Symposium, Rocquencourt, June 17–21, 1974. Edited by A. Bensoussan and J. L. Lions. VIII, 757 pages. 1975.

Vol. 108: F. Bauer et al., Supercritical Wing Sections II. A Handbook. V, 296 pages. 1975.

Vol. 109: R. von Randow, Introduction to the Theory of Matroids. IX, 102 pages. 1975.

Vol. 110: C. Striebel, Optimal Control of Discrete Time Stochastic Systems. III. 208 pages. 1975.

Vol. 111: Variable Structure Systems with Application to Economics and Biology. Proceedings 1974. Edited by A. Ruberti and R. R. Mohler. VI, 321 pages. 1975.

Vol. 112: J. Wilhlem, Objectives and Multi-Objective Decision Making Under Uncertainty. IV, 111 pages. 1975.

Vol. 113: G. A. Aschinger, Stabilitätsaussagen über Klassen von Matrizen mit verschwindenden Zeilensummen. V, 102 Seiten. 1975.

Vol. 114: G. Uebe, Produktionstheorie. XVII, 301 Seiten. 1976.

Vol. 115: Anderson et al., Foundations of System Theory: Finitary and Infinitary Conditions. VII, 93 pages. 1976

Vol. 116: K. Miyazawa, Input-Output Analysis and the Structure of Income Distribution. IX, 135 pages. 1976.

Vol. 117: Optimization and Operations Research. Proceedings 1975. Edited by W. Oettli and K. Ritter. IV, 316 pages. 1976.

Vol. 118: Traffic Equilibrium Methods, Proceedings 1974. Edited by M. A. Florian. XXIII, 432 pages. 1976.

Vol. 119: Inflation in Small Countries. Proceedings 1974. Edited by H. Frisch. VI, 356 pages. 1976.

Vol. 120: G. Hasenkamp, Specification and Estimation of Multiple-Output Production Functions. VII, 151 pages. 1976.

Vol. 121: J. W. Cohen, On Regenerative Processes in Queueing Theory. IX, 93 pages. 1976.

Vol. 122: M. S. Bazaraa, and C. M. Shetty,Foundations of Optimization VI. 193 pages. 1976

Vol. 123: Multiple Criteria Decision Making. Kyoto 1975. Edited by M. Zeleny. XXVII, 345 pages. 1976.

Vol. 124: M. J. Todd. The Computation of Fixed Points and Applications. VII, 129 pages. 1976.

Vol. 125: Karl C. Mosler. Optimale Transportnetze. Zur Bestimmung ihres kostengünstigsten Standorts bei gegebener Nachfrage. VI, 142 Seiten. 1976.

Vol. 126: Energy, Regional Science and Public Policy. Energy and Environment I. Proceedings 1975. Edited by M. Chatterji and P. Van Rompuy. VIII, 316 pages. 1976.

Vol. 127: Environment, Regional Science and Interregional Modeling. Energy and Environment II. Proceedings 1975. Edited by M. Chatterji and P. Van Rompuy. IX, 211 pages. 1976.

Vol. 128: Integer Programming and Related Areas. A Classified Bibliography. Edited by C. Kastning. XII, 495 pages. 1976.

Vol. 129: H.-J. Lüthi, Komplementaritäts- und Fixpunktalgorithmen in der mathematischen Programmierung. Spieltheorie und Ökonomie. VII, 145 Seiten. 1976.

Vol. 130: Multiple Criteria Decision Making, Jouy-en-Josas, France. Proceedings 1975. Edited by H. Thiriez and S. Zionts. VI, 409 pages. 1976.

Vol. 131: Mathematical Systems Theory. Proceedings 1975. Edited by G. Marchesini and S. K. Mitter. X, 408 pages. 1976.

Vol. 132: U. H. Funke, Mathematical Models in Marketing. A Collection of Abstracts. XX, 514 pages. 1976.

Vol. 133: Warsaw Fall Seminars in Mathematical Economics 1975. Edited by M. W. Loś, J. Loś, and A. Wieczorek. V. 159 pages. 1976.

Vol. 134: Computing Methods in Applied Sciences and Engineering. Proceedings 1975. VIII, 390 pages. 1976.

Vol. 135: H. Haga, A Disequilibrium – Equilibrium Model with Money and Bonds. A Keynesian – Walrasian Synthesis. VI, 119 pages. 1976.

Vol. 136: E. Kofler und G. Menges, Entscheidungen bei unvollständiger Information. XII, 357 Seiten. 1976.

Vol. 137: R. Wets, Grundlagen Konvexer Optimierung. VI, 146 Seiten. 1976.

Vol. 138: K. Okuguchi, Expectations and Stability in Oligopoly Models. VI, 103 pages. 1976.

Vol. 139: Production Theory and Its Applications. Proceedings. Edited by H. Albach and G. Bergendahl. VIII, 193 pages. 1977.

Vol. 140: W. Eichhorn and J. Voeller, Theory of the Price Index. Fisher's Test Approach and Generalizations. VII, 95 pages. 1976.

Vol. 141: Mathematical Economics and Game Theory. Essays in Honor of Oskar Morgenstern. Edited by R. Henn and O. Moeschlin. XIV, 703 pages. 1977.

Vol. 142: J. S. Lane, On Optimal Population Paths. V, 123 pages. 1977.

Vol. 143: B. Näslund, An Analysis of Economic Size Distributions. XV, 100 pages. 1977.

Vol. 144: Convex Analysis and Its Applications. Proceedings 1976. Edited by A. Auslender. VI, 219 pages. 1977.

Vol. 145: J. Rosenmüller, Extreme Games and Their Solutions. IV, 126 pages. 1977

Vol. 146: In Search of Economic Indicators. Edited by W. H. Strigel. XVI, 198 pages. 1977.

Vol. 147: Resource Allocation and Division of Space. Proceedings. Edited by T. Fujii and R. Sato. VIII, 184 pages. 1977.

Vol. 148: C. E. Mandl, Simulationstechnik und Simulationsmodelle in den Sozial- und Wirtschaftswissenschaften. IX, 173 Seiten. 1977.

Vol. 149: Stationäre und schrumpfende Bevölkerungen: Demographisches Null- und Negativwachstum in Österreich. Herausgegeben von G. Feichtinger. VI, 262 Seiten. 1977.

Vol. 150: Bauer et al., Supercritical Wing Sections III. VI, 179 pages. 1977.

Vol. 151: C. A. Schneeweiß, Inventory-Production Theory. VI, 116 pages. 1977.

Vol. 152: Kirsch et al., Notwendige Optimalitätsbedingungen und ihre Anwendung. VI, 157 Seiten. 1978.

Vol. 153: Kombinatorische Entscheidungsprobleme: Methoden und Anwendungen. Herausgegeben von T. M. Liebling und M. Rössler. VIII, 206 Seiten. 1978.

Vol. 154: Problems and Instruments of Business Cycle Analysis. Proceedings 1977. Edited by W. H. Strigel. VI, 442 pages. 1978.

Vol. 155: Multiple Criteria Problem Solving. Proceedings 1977. Edited by S. Zionts. VIII, 567 pages. 1978.

Vol. 156: B. Näslund and B. Sellstedt, Neo-Ricardian Theory. With Applications to Some Current Economic Problems. VI, 165 pages. 1978.

Vol. 157: Optimization and Operations Research. Proceedings 1977. Edited by R. Henn, B. Korte, and W. Oettli. VI, 270 pages. 1978.

Vol. 158: L. J. Cherene, Set Valued Dynamical Systems and Economic Flow. VIII, 83 pages. 1978.

Vol. 159: Some Aspects of the Foundations of General Equilibrium Theory: The Posthumous Papers of Peter J. Kalman. Edited by J. Green. VI, 167 pages. 1978.

Vol. 160: Integer Programming and Related Areas. A Classified Bibliography. Edited by D. Hausmann. XIV, 314 pages. 1978.